Compact Textbooks in Mathematics

This textbook series presents concise introductions to current topics in mathematics and mainly addresses advanced undergraduates and master students. The concept is to offer small books covering subject matter equivalent to 2- or 3-hour lectures or seminars which are also suitable for self-study. The books provide students and teachers with new perspectives and novel approaches. They may feature examples and exercises to illustrate key concepts and applications of the theoretical contents. The series also includes textbooks specifically speaking to the needs of students from other disciplines such as physics, computer science, engineering, life sciences, finance.

- **compact:** small books presenting the relevant knowledge
- **learning made easy:** examples and exercises illustrate the application of the contents
- **useful for lecturers:** each title can serve as basis and guideline for a semester course/lecture/seminar of 2-3 hours per week.

Christopher Baltus

Geometry by Its Transformations

Lessons Centered on the History from 1800-1855

 Birkhäuser

Christopher Baltus
Department of Mathematics
State University of New York at Oswego
Oswego, NY, USA

ISSN 2296-4568   ISSN 2296-455X   (electronic)
Compact Textbooks in Mathematics
ISBN 978-3-031-72280-6   ISBN 978-3-031-72281-3   (eBook)
https://doi.org/10.1007/978-3-031-72281-3

Mathematics Subject Classification: 01A55, 51-03, 51-01, 97G50

This book is published under the imprint Birkhäuser, www.birkhauser-science.com by the registered company Springer Nature Switzerland AG
The registered company address is: Gewerbestrasse 11, 6330 Cham, Switzerland

If disposing of this product, please recycle the paper.

Preface

Although this book was written in 2022 and 2023, it is the culmination of a long project. As an education student at the University of Chicago in the 1970s, I encountered ancient Greek geometry, the topic of the first chapter, and was introduced, by Zalman Usiskin, to *transformation geometry*, the topic of the last chapter. The development of the intervening chapters was supported by the college where I taught, SUNY Oswego, with its Penfield Library, and with the kind encouragement and suggestions from many colleagues. My thanks to members of the Canadian Society for History and Philosophy of Mathematics, among whom the late Hardy Grant was especially helpful. I recognize the internet sites which bring a library of historical mathematics texts to an office or home desktop computer, notably bnf.fr, the *Bibliothèque nationale de France*; the University of Michigan's quod.lib.umich.edu; and the *Internet Archive*. Most recently, I have had access to the library of Vassar College and the support of the publisher Birkhäuser, with editor Frida Trotter. My thanks to all, and to my wife, Banna Rubinow, for her steady encouragement and ruthless editing.

Poughkeepsie, NY, USA   Christopher Baltus

Introduction

What is geometry? It is a simple question to which there is no simple answer. In this book, our approach is to examine an aspect of geometry, *transformations*, by its history, emphasizing the period of years 1800–1855.

A simple answer which has a good measure of truth is that, until 1800, geometry consisted of elaboration on *The Elements* of Euclid, written about 300 BCE. Most works of geometry followed the example of *The Elements* in starting with assumed statements and definitions—not always listed explicitly—followed by propositions whose justification was based on the assumed statements and already proven propositions.

However, putting the weight for the development of geometry on Euclid misses some important aspects of geometry and its history. Euclid produced almost no numerical answers. For example, he showed in Book 12 Prop. 1 and 2 that the area of a circle of radius r is proportional to r^2, but he gave no numerical approximation for the ratio of those two quantities, the ratio we call π. Another mathematician working within Greek culture, Archimedes, showed, a long generation later, that π as noted above is also the ratio of the circumference of a circle to its diameter, and that the value is between 3 10/71 and 3 1/7. Further, those bounds could be sharpened by the same method. In the third century CE, Lui Hui, in China, achieved, independently, essentially the same result. (See Application 4 of Chapter 2.)

A significant topic absent from *The Elements* is spherical geometry. It was developed by Greek mathematicians in the centuries after Euclid, and was further advanced by Islamic mathematicians in their medieval period. Trigonometry grew with spherical geometry, at the hands of both ancient Greek and Islamic mathematicians. But neither the discovery of numerical values nor spherical geometry nor trigonometry was important in the blossoming of geometry in the early nineteenth century. The interested reader will find a historical introduction to spherical geometry in [101] and a detailed history of trigonometry up to year 1550 in [100].

In the seventeenth century, the introduction of the cartesian plane and coordinate geometry, most notably by René Descartes in 1637, offered a new approach to geometry. Think of a line, perhaps the most important element of geometry. Euclid included *line* in his defined terms, but he gave not so much a definition as a description, aimed at readers who already have the concept: "A *line* is breadthless length" [39]. In coordinate geometry, on the other hand, a line is the set of points (x, y) satisfying an equation of the form $ax + by + c = 0$ when a and b are not both 0. In the second half of the seventeenth century, Isaac Newton used all four quadrants of the cartesian plane as we do now, with none of the hesitancy shown by Descartes about negative numbers. Soon after, there came the flood of books and papers handling the calculus and various topics in geometry by coordinate algebra, a flood that continues to this day. This geometry is called *analytic*. Geometry in the style of Euclid did not disappear but rather continued as an almost separate subject, which we have come to call *synthetic geometry*. Notably, Newton chose to present his master work, the *Principia Mathematica*, of 1687, as synthetic geometry.

It is worth noting that a work of true projective geometry, the *Brouillon project* [35], by Girard Desargues (1591–1661), appeared in 1639 and would be followed a generation later by a short work of Philippe de la Hire (1640–1718) [60] that employed projective methods. Both works were essentially unknown in 1800, and would remain largely unknown until their subject, projective geometry, was developed in the following half century. It seems fair to conclude that other seventeenth-century geometers were not ready for this projective approach to their subject. The nineteenth-century development of projective methods will be treated in several chapters of this book.

This book works, with a few exceptions, in synthetic geometry.

Why 1800 to 1855?

Synthetic geometry in 1800 and in 1855 were very different enterprises. The synthetic geometry of 1800, perhaps best exemplified by Adrien-Marie Legendre's *Élémens de Géométrie* [63] of 1794, would have felt familiar to Greek geometers of two millennia earlier. In fact, Legendre's *Géométrie* was largely a revisiting of Euclid's *Elements*, in the style of Euclid, to which were added sections on trigonometry, on area and volume of figures, and a few other topics, such as a proof that π is irrational. In that style, a figure is given, with certain specific features about segment length or parallel lines, and so on, and one is to demonstrate that some other feature necessarily holds. In the six decades after 1794, various ways were introduced to transform a diagram, as part of a transformation of the entire plane, so one might prove a claim in the transformed figure, thus justifying the original claim. Attention moved to those transformations themselves. Transformations were not entirely new: an artist transforms a floor pattern to the plane of the canvas on which the painting appears. Isaac Newton gave a method of transforming curves in Book 1 Lemma 22 of his *Principia Mathematica*, of 1687. Leonhard Euler (1707–1783) had related figures on a plane and in space by what he called "similitude." But never before was transformation so central to the study of geometry. In 1827, A. F. Moebius gave an overview of transformations, in his *Der Barycentrische Calcul*, listing four different transformations. In this sense, he made transformations, in themselves, a subject of study.

With different transformations came attention to different planes in which we do geometry. Initially, these new planes were extensions of the familiar *real plane*. The real plane, denote as $\mathbb{R}^2$, is the set of ordered pairs (x, y), where x and y range over all the real numbers. When mathematicians decided that parallel lines could meet, we needed to add to the real plane a *horizon line* of *points at infinity*, at which these parallels meet. This gave us the *real projective plane*, $\mathbb{P}^2$. Moebius, in 1827, improved by Julius Plücker in 1831, gave us a precise way to define $\mathbb{P}^2$ using ordered triples of real numbers.

In $\mathbb{P}^2$ there are no parallel lines; all lines meet. So to make sense of the transformation that Moebius called an *affinity*, where the plane is transformed in a way that parallel lines are mapped to parallel lines, we need a plane with parallels. Thus, $\mathbb{R}^2$ was to be referred to as an *affine plane*, where we specifically require that for each line m and point X not on m, there is a line parallel to m. This is a strange but not rare phenomenon where we use a concept, such as an affine plane, long before we define it. We'll also see the plane where we add just one point at infinity to $\mathbb{R}^2$, where all lines lie on this point; the *inversion* transformation operates in this plane. (In this book, we concentrate on planes, which are two-dimensional spaces, and only occasionally venture off that plane into three-dimensional space.)

After Adrien-Marie Legendre's *Géométrie*, of 1794, closing the period dominated by the *Elements* of Euclid, Gaspard Monge's *descriptive geometry*, in lessons written in 1797 for the new École Polytechnique [71], and Lazare Carnot's *Géométrie de position*, of 1803, pointed to developments to come. At the far end of our period, we see Michel Chasles's *Traité de Géométrie Supérieure*, of 1852 [27], and George Salmon's *A Treatise of Conic Sections* [91] of 1848 and 1855, as summaries of geometry at the close of an era, well-written works that would continue to instruct students for decades but barely breaking new ground. Between these dates we concentrate, in sequence, on work of Charles-Julien Brianchon, of Jean-Victor Poncelet, of August Ferdinand Moebius, of Jacob Steiner, and of Karl Georg Christian von Staudt. Von Staudt's *Geometrie der Lage*, of 1847, was to become, several decades after its composition, a very influential work for the new era in projective geometry, but at the same time rested on what had been found in the previous 30 years. By contrast, von Staudt's *Beiträge zur Geometrie der Lage* [105], which appeared in three volumes from 1856 to 1860, is a very different work, as much for the abstract tone as for the advanced content. There von Staudt introduced a complex projective space and an algebra derived from geometric axioms, topics in the emerging newer geometry.

We have a short introduction to *inversion geometry*, concentrating on an 1836 work of Giusto Bellavitis. The last of the transformations we consider, the *Moebius transform*, was developed in papers by A. F. Moebius in 1853 and 1855. It was in line with the transformations developed since 1820, but, in working in the complex plane, pointed to future work. A further reason for selecting 1855 to date the end of our era is the 1854 inaugural lecture at the University of Göttingen by Bernhard Riemann (1826–1866) [90]. This lecture was the first step toward the new view of geometry that would dominate the last third of the nineteenth century. The impact was not immediate; Riemann's talk would only become a published paper in 1868, after Riemann's death, and was soon followed by an article that spread Riemann's ideas, "On the facts which lie at the foundations of geometry," by Hermann von Helmholtz [54]. A plane was generalized to a "surface," which in turn was generalized as a two-dimensional "manifold," and the continuous movement of an n-manifold generates an $n + 1$ manifold.

A word in the book title deserves mention: *transformation*. We now think of a *transformation* as a one-to-one function mapping a plane or space onto itself. Significantly, in the first study of transformations as a general topic, A. F. Moebius's *Barycentrische Calcul* of 1827, the term *Verwandtschaft*, meaning "relationship," was used in place of our name *transformation*. Although the name *function* was widely recognized in mathematics since Euler's 1748 *Introduction to analysis of the infinite*, it had kept the context in which Euler had used it: a function of one or several variables was an mathematical expression in those variables. Many decades would pass before we would speak of a function of geometric objects. The relevance of this observation will arise with the first transformation we examine, the *dilation*. As a function, the dilation with center at the origin and non-zero scale factor k maps each point (x, y) to (kx, ky). However, the first use of dilation as a named feature of mathematics was as a *similitude*, by Euler and Poncelet. It was a pairing of similar figures. For Poncelet, the similar figures involved were paired point by point so that corresponding segments were parallel. Euler's *similitude* paired any similar figures with the same orientation. A *function* is a one-way street, pairing an *input* item with its *image*, while a *similitude* was a two-way street, a *pairing* or a *relationship*, as Moebius expressed it.

After a brief note, we turn to a selective summary of ancient Greek mathematics, with particular attention to Euclid's *Elements*. This chapter is to get the reader up to speed on the background expected throughout the book.

Note to Reader

Any book is written with a certain reader in mind. Hopefully, this book will be read by college mathematics majors and college instructors, but any reader with a good high school geometry background and enthusiasm for mathematical reasoning should be able to work through it. It could serve as a text for the course generally called "College Geometry," if one wants geometry firmly situated in its history.

We are selective with historical details, favoring detailed exposition about fewer ideas and propositions, and concentrating on rather few mathematicians. After the chapter on ancient Greek mathematics, we will examine the simplest of the geometric transformations, the *dilation*. In the chapter on dilations, we point out features which characterize a geometric transformation.

The biographer Lytton Strachey began his 1918 *Eminent Victorians*, "The history of the Victorian Age will never be written: we know too much about it." [99] One could say the same thing about nineteenth-century mathematics, which exploded in its range, especially after 1825. In this short book, any claim for completeness would be ludicrous, and I, the author, did not try.

A true history would pay much attention to what the contemporary actors found most important. I have violated that principle by ignoring many of the contemporary actors and many of their concerns. Consider Jean-Victor Poncelet (1788–1867), a founder of projective geometry and a major subject of this book. He was very interested in imaginary points and imaginary lines, and Poncelet wrote several papers, such as [88], on polar reciprocals; he based his claims for priority in the introduction of *duality* on his work with polar reciprocals. This book has nothing on polar reciprocals and very little on imaginary points or lines. We barely mention surfaces and transformations in three-dimensional space.

This work has the uncommon goal of combining history and mathematical exposition of the basic ideas. One guiding idea is that it should work as an introductory geometry course for college mathematics majors, where the mathematics is placed in its historical context. A principle in choosing what to omit is that we should focus on what has endured in importance.

Further, although the title suggests that only work from 1800 to 1855 is treated, for mathematical coherence I went outside those boundaries where appropriate. This book ends with material from the years after 1855, in a short chapter on a topic of non-Euclidean geometry, and a chapter on the introduction of geometric transformations into school mathematics, especially around 1900 in France.

Many chapters end with one or several *Applications*. A reader interested in a spare history can generally skip those. And there are *Exercises* at the end of most chapters, the sort of thing one expects in mathematical exposition; the reader may skim these without loss of historical continuity but with the caution that some significant content appears only in Exercises. An appendix has solutions and hints for most of the Exercises. One can take the book for a fairly slim history or for a meatier course in geometry.

This book pays particular attention to straightedge-compass constructions; constructions often call on the dilation or homology or inversion transformation. In an Application at the end of the Inversion chapter, we review the ten construction problems that ask for a circle tangent to three given figures, which can be points or lines or circles.

About matrices. There was work on matrix mathematics throughout the nineteenth century, including the theory of determinants and calculations with matrices, but only in the twentieth century did it become a standard tool in geometry. Nevertheless, matrices are especially helpful in understanding geometric transformations and carrying out related computations. So along with descriptions of the work of the nineteenth-century mathematicians, which is faithful to the original works, I have included analysis in terms of matrices. With this in mind, the book includes an appendix setting out the matrix mathematics needed to follow what is in the book. In the special case of Moebius's 1827 work, the matrices here presented are just a simpler representation of systems of equations in the original; they do not distort Moebius's achievement.

Contents

Greek Background

1   Forgotten Mathematics

European mathematicians in 1800 were ignorant of much of the geometry developed after 1600. We have already noted one work that was lost, Girard Desargues' *Brouillon project*, of 1639, although some content was known from the negative critiques which did survive. A transcription by Philippe de la Hire would be discovered in 1845. Other works were forgotten, although not completely lost. This group includes Desargues' Theorem of 1648, that triangles in perspective from a point have corresponding sides that meet in collinear points. Also forgotten was Pascal's Theorem, first published in 1640, that if a hexagon is inscribed in a conic section, the opposites sides of the hexagon meet in collinear points.

There was a similar story for Ceva's Theorem: Given lines on the three vertices of a triangle ABC, meeting opposite sides, respectively, in a, b, and c, the three lines are concurrent exactly when

$$aB \cdot bC \cdot cA = aC \cdot bA \cdot cB.$$

There seem to be two independent sources that were forgotten by 1800: Giovanni Ceva's in *De lineis rectis*, of 1678 [25, 26], and Johann Bernoulli's [12, p. 33], of 1742. Ceva's proof involved statics, weights placed at points in a plane. We do not see the theorem referred to again until Carnot's *Géométrie de Position*, of 1803, and his *Essai sur la théorie des transversals*, of 1806. Neither mentions Ceva or Bernoulli. In recent years, fragments of a work by an eleventh century king of Muslim Zaragoza, Yusuf al-Mu'taman, have been found to include Ceva's Theorem. We will see a different proof later in this chapter.

What we can securely assume in the geometry background of mathematicians working around 1800 is a collection of Greek classics, many published in Latin in the late 1500s, if not earlier. The first is Euclid's *Elements*, from about 300 BCE. Next is the *Conics* of Apollonius, from about 200 BCE. Then we have Ptolemy's

© The Author(s), under exclusive license to Springer Nature Switzerland AG 2025
C. Baltus, *Geometry by Its Transformations*, Compact Textbooks in Mathematics,
https://doi.org/10.1007/978-3-031-72281-3_1

Almagest and his *Planisphere*, from the second century CE. Finally, there was Pappus's *Mathematical Collection*, from the fourth century CE. We will present here the parts of these works that are most important to follow developments in the first half of the nineteenth century.

2   Euclid

The *Elements* of Euclid, written in Alexandria, Egypt, is the earliest comprehensive mathematics text that has survived into modern times. Proclus, in the fifth century CE, wrote that Euclid "brought together the Elements, systematizing many of the theorems of Eudoxus, perfecting many of those of Theatetus, and putting in irrefutable demonstrable form propositions that had been rather loosely established by his predecessors" [39]. It is written in thirteen "books." It served, with modification, as the primary introduction to geometry for students well into the twentieth century. Not only did it present the foundational propositions of geometry, but it also established the structure of a rigorous mathematical presentation: we begin with definitions and the axioms—statements to be taken as true—to be followed by propositions proved making use of the axioms and already proven propositions.

3   The Geometry of Euclid's *Elements*: Book 1

Book 1 begins with *Definitions*. Then, following the form recommended by Aristotle, are the *Postulates* and *Common Notions*, followed by *Propositions*. Postulates are statements to be assumed that are specific to a particular science, such as mathematics. Common Notions are assumed statements common to all sciences, such as

Common Notion 1 "Things equal to the same thing are equal to each other."

Also helpful is

Common Notion 5 "The whole is greater than its part." [39]

Common Notion 5 is applicable to angles with a common vertex and common side, where one is inside the other. Based on that Common Notion, we can declare that the inside angle is less in measure than the outside angle. See Fig. 1.1 Left where, based on Common Notion 5, angle *DBC* has lesser measure than angle *ABC*. Modern secondary textbooks may label that claim as the *Angle Measure Inequality*.

These, in modern language, are the definitions in Euclid's *Elements* of most interest to us.

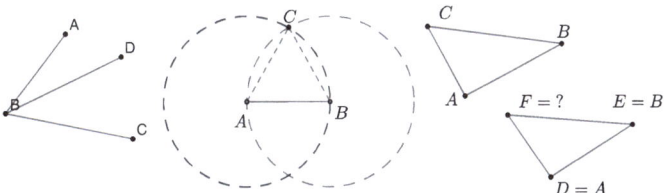

Fig. 1.1 Left: $m\angle DBC < m\angle ABC$. Center: Construction of equilateral triangle on base AB: Euclid Book 1 Prop. 1. Right: Proof of SAS triangle congruence: Euclid Book 1 Prop. 4

Definition 10 If two lines meet so that adjacent angles are equal [congruent], then the lines are called *perpendicular* and the angles formed are *right angles*. (This definition led Euclid to include as the fourth postulate: "All right angles are equal." This guarantees that angles produced by perpendicular lines at different positions in the plane are congruent to each other.)

Definition 15 A *circle* is designated by its center and radius; it is the set of points whose distance from the center equals the given radius.

Definition 23 Lines that lie in one plane and never meet, even if produced indefinitely in both directions, are called *parallel*.

There are five **Postulates** of Book 1.

1. To draw a line [segment] joining any two points.
2. To extend any line segment.
3. To draw a circle with any given center and radius.
4. All right angles are equal [congruent].
5. (Parallel Postulate) If a transversal cuts two lines so that the two interior angles on one side of the transversal are together less than a straight angle, then the two lines meet on the side with those interior angles.

The propositions of Book 1 begin with

Proposition 1 To construct an equilateral triangle on a given base.

See the construction of Fig. 1.1 Center.
Book 1 includes the triangle congruence theorems:

Prop. 4 *SAS*,
Prop. 8 *SSS*,
Prop. 26 *ASA* and *AAS*.

The expression SAS is the theorem that when the vertices of two triangles correspond so that two sides and the included angle of one agree with the corresponding parts of the second triangle, then the triangles are congruent. It has an interesting proof, which involves a transformation.

Proof Book 1 Prop. 4. Let $\triangle ABC$ and $\triangle DEF$ have $AB = DE$, $AC = DF$ and $\angle A \cong \angle D$. See Fig. 1.1 Right. We show that $\triangle ABC$ can be made to coincide with $\triangle DEF$. We superimpose $\triangle ABC$ on $\triangle DEF$ so A lies on D, side AB on DE, and C on the same side of line DE as F. Because $AB = DE$, then B coincides with E. Because angle BAC is congruent to angle EDF, then side AC lies on side DF. But $AC = DF$ so C coincides with F. Thus, side BC coincides with EF. So the triangles coincide. Euclid continued in the proposition to state that the corresponding angles are congruent.   □

In school geometry, we take Propositions 4, 8, and 26 as simply claims that triangles satisfying the given conditions are *congruent*, and then conclude that other corresponding sides and angles are congruent, by definition of congruence. It is common to write $CPCTC$, "corresponding parts of congruent triangles are congruent."

In modern school vocabulary, line segments and angles have real number *measures*, and when two segments or angles agree in measure they are called *congruent*. Polygons are *congruent*, with a certain correspondence of vertices, when corresponding sides and corresponding angles are congruent. *Equal* is reserved for numbers, such as segment lengths, or two names for the same figure. The distinction between "congruent" and "equal" appeared in the twentieth century since, in set language, sets of points are only "equal" when they are formed of exactly the same points. We will generally use the term "congruent," but will occasionally use "equal" when a work considered used that term. The symbol $\cong$ denotes congruence.

It is now standard in school geometry to distinguish sets of points, such as segments and angles, from their measures. A *segment* is the part of a line between two points, together with those two points. The segment whose endpoints are A and B is denoted with a bar, as $\overline{AB}$; the *ray*, or *halfline*, whose endpoint is A and lies on B is $\overrightarrow{AB}$. The line on A and B, of indefinite length in both directions is $\overleftrightarrow{AB}$. To indicate the number which is the measure of a segment, we just give the endpoints, as AB. For the measure of an angle, we place m in front of the angle symbol. We shall follow the custom widely observed outside school mathematics books of giving just two points of a line or the two endpoints of a segment, where wording or context indicates whether a segment or a line or its measure is intended. Note the symbols used in the proof of Euclid's Proposition 4. We write $AB = DE$

Fig. 1.2 Euclid's
justification of angle bisector
construction Book 1 Prop. 9

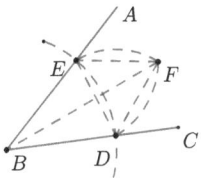

since the lengths of segments $\overline{AB}$ and $\overline{DE}$ are equal, but we write $\angle A \cong \angle D$ since the angles have equal measure but are not the same sets of points.

In Euclid's geometry, *line* was what we now call a *segment*, but subject to extension. Ancient Greeks would not speak of a line as infinitely long, or even indefinitely long.

Prop. 5 and 6 are the *Isosceles Triangle Theorem*: Two sides of a triangle are congruent exactly when the opposite angles are congruent.

Propositions 9 to 12, and 23, provide the procedure—and the justification—for these constructions:

Prop. 9 To bisect an angle.
Prop. 10 To bisect a segment.
Prop. 11 and 12 To construct a perpendicular to a given line on a given point.
Prop. 23 To construct an angle congruent to a given angle.

All these constructions call on the SSS Triangle Congruence Theorem in their justifications. A proof of the SSS Congruence Theorem, in one case, is an Exercise at the end of this chapter.

Let us look at Euclid's proof of Prop. 9, to bisect a given angle, $\angle ABC$. See Fig. 1.2. First select a point, D, not B, on side BC. By Postulate 3, we draw the circle with center B and lying on D, and meeting side BA at a point E. By Postulate 1, we draw segment DE. Then by Proposition 1, we construct an equilateral triangle on base DE, whose third vertex is F. Draw BF. Then triangles BFE and BFD are congruent by SSS. (Side BF corresponds to itself.) Then, by $CPCTC$, $\angle EBF \cong \angle DBF$. $\square$

We now turn to a sequence of propositions concerning angles, parallel lines, and, in the pivotal Prop. 16, an exterior angle of a triangle. (The term "linear pair" is found in some secondary texts; it was not used by Euclid.)

Prop. 13 Angles that form a *linear pair* are supplementary.

In Fig. 1.3 Top Left, angle DBA and angle DBC form a *linear pair* since they are adjacent angles whose noncommon sides are opposite rays, i.e., form a line.

Prop. 15 Vertical angles are congruent.

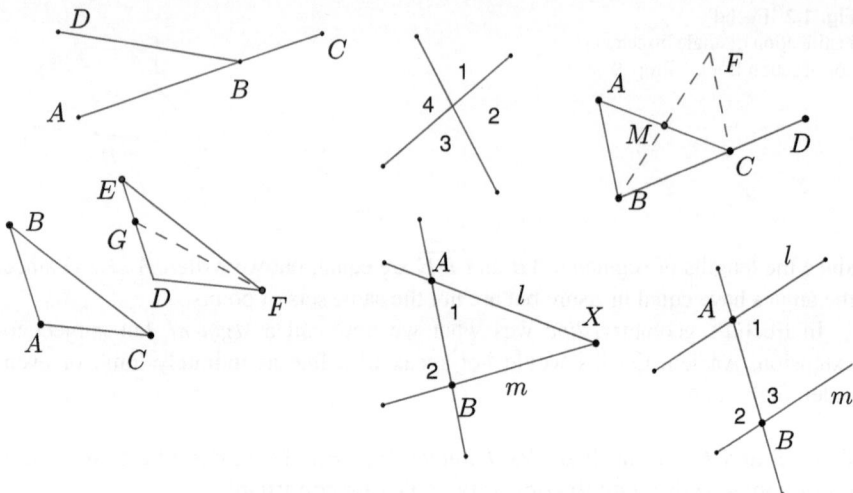

Fig. 1.3 Diagrams, Euclid's *Elements* Book 1. Top Left: Angles ABD and DBC are a Linear Pair. Top Center: Prop. 15, Vertical angles are congruent. Top Right: Prop. 16. Lower Left: Prop. 26 AAS Congruence. Lower Middle: Prop. 27. Lower Right: Prop. 29

In Fig. 1.3 Top Center, angles 1 and 3, and angles 2 and 4, are pairs of congruent vertical angles. Angles 1 and 3, for example, are each supplementary to angle 2, so they are congruent to each other.

Prop. 16 *Euclid's Exterior Angle Theorem.* An exterior angle of a triangle is greater in measure than either opposite interior angle.

Proof Proposition 16. See Fig. 1.3 Upper Right. With $\triangle ABC$, let side BC be extended past C to D, creating exterior angle ACD. Let M be the midpoint of AC (M can be constructed by Prop. 10), then draw BM (Post. 1: Two points can be joined by a line segment.) and extend it past M (Post. 2: Any line segment can be extended.), marking F on this extension so $BM = MF$ (Post. 3: To draw a circle with a given center and radius.). Join F and C. Then $\angle BMA \cong \angle FMC$ (vertical angles). By SAS, $\triangle AMB \cong \triangle CMF$. So, as corresponding parts of congruent triangles (part of Prop. 4), $\angle A \cong \angle ACF$. $m\angle ACD > m\angle ACF$ (Common Notion 5). It follows that exterior angle ACD is greater than opposite interior angle A. □

To see Euclid's use of indirect proof (proof by contradiction), here is the proof of the AAS part of Proposition 26. The proof calls on Prop. 16. See Fig. 1.3 Lower Left. We are given $\triangle ABC$ and $\triangle DEF$ where $\angle A \cong \angle D$, $\angle B \cong \angle E$, and $AC = DF$. The strategy is to show that AB equals DE by supposing, first, that $AB < DE$ and deriving a contradiction. (The case $AB > DE$ would be handled in a similar way.)

Assuming $AB < DE$, mark G on DE so $DG = AB$. We draw GF, and note that $\triangle DGF \cong \triangle ABC$ (SAS). With, then, $\angle DGF \cong \angle B$, we have $\angle DGF \cong$

$\angle E$, which violates Euclid's Exterior Angle Theorem. Therefore, $AB = DE$, so $\triangle ABC \cong \triangle DEF$ (SAS). $\square$

There is a crucial foundational distinction between Propositions 16 and 32:

Prop. 32 An exterior angle of a triangle is equal in measure to the sum of the two opposite interior angles.

Let us recall Postulate 5, the *Parallel Postulate*:

Postulate 5 If a transversal cuts two lines so that the two interior angles on one side of the transversal are together less than a straight angle, then the two lines meet on the side with those interior angles.

As we have seen, Prop. 16 can be proved without invoking Postulate 5, the *Parallel Postulate*. The stronger Prop. 32 requires it. There is a corresponding distinction between Prop. 17, that any two angles of a triangle are together less than two right angles, and a second part of Prop. 32, that the three angles of a triangle are together equal to two right angles. And there is yet another corresponding distinction between Prop. 27, that two lines cut by a transversal so alternate interior angle are congruent must be parallel, and its converse, Prop. 29. In each pair, the first proposition can be proved without the Parallel Postulate while the second proposition requires the Parallel Postulate. The geometry developed without any postulate about the existence of parallels is called *Absolute Geometry*. Euclid's Propositions 1 through 28 of Book 1 are part of Absolute Geometry. It was proved in the nineteenth century that the second proposition of each of these pairs requires the Parallel Postulate or an equivalent assumption. (In Chapter 13, we outline one of these proofs.)

Here are Propositions 27 and 28, part of Absolute Geometry:

Prop. 27 *Alternate Interior Angle Theorem.* If two lines are cut so alternate interior angles are congruent, then the lines are parallel. (In Fig. 1.3 Lower Right, angles 1 and 2 are alternate interior angles.)

Likewise,

Prop. 28 If corresponding angles are congruent, then the lines are parallel.

Proof *Alternate Interior Angle Theorem*, Proposition 27. See Fig. 1.3 Lower Center. We suppose l and m are cut at A and B, respectively, in such a way that alternate interior angles 1 and 2 are congruent. Further, we suppose l and m are not parallel, meeting at a point X. But then exterior angle 2 equals in measure opposite interior angle 1 of $\triangle ABX$, violating Euclid's Exterior Angle Theorem. We conclude that l and m are parallel. $\square$

Beginning with Prop. 29, Euclid presented propositions whose proofs require the Parallel Postulate.

Prop. 29 *Alternate Interior Angle Theorem. Converse of Prop. 27.*

If parallel lines are cut by a transversal, then alternate interior angles (and corresponding angles) are congruent.

Prop. 32 An exterior angle of a triangle equals, in measure, the sum of the two opposite interior angles, and the interior angles of a triangle are, together, equal to two right angles.

Prop. 33 Opposite sides of a parallelogram are congruent. (A *parallelogram* is defined to be a quadrilateral where both pairs of opposite sides are parallel.)

Proof (Proposition 29) See Fig. 1.3 Lower Right. Suppose l and m are parallel, but are cut by line AB so $m\angle 2 > m\angle 1$. Now, the sum $m\angle 2 + m\angle 3$ equals two right angles (Linear Pair), so $m\angle 3 + m\angle 1$ is less than two right angles. By Post. 5, l and m meet, contradicting what was supposed. Therefore, $\angle 2 \cong \angle 1$. □

Book 1 ends with the Pythagorean Theorem and its converse, in Propositions 47 and 48.

Proofs of Propositions 32 and 33 are in the Exercises at the end of the chapter.

4 The Geometry of Euclid's *Elements*: Triangle Similarity and Circles in Books 6 and 3

Although similar triangles are introduced only in Book 6, it will help to give a definition and several theorems from Book 6 which we will apply in considering Book 3, on circles.

Definition
(Book 6, Def. 1) Two polygons are *similar* if there is a correspondence of vertices so corresponding angles are congruent and corresponding sides are in proportion.

Book 6 Propositions 4, 5, 6
Triangles are similar when there is a correspondence of vertices so that

AA: two pairs of corresponding angles are congruent, or
SSS: the three pairs of corresponding sides are in proportion, or
SAS: two pairs of corresponding sides are in proportion and the included angles congruent.

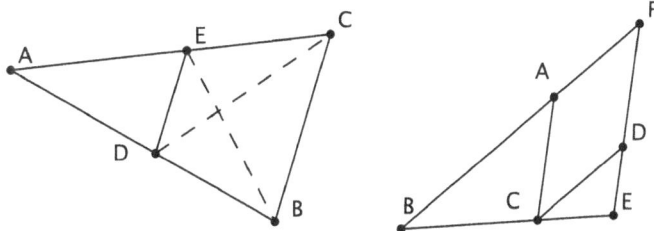

Fig. 1.4 Euclid Book 6. Left: Prop. 2, Side Splitter Theorem. Right: Prop. 4, Proof of the AA Similarity Theorem

The first proposition of Book 6 tells us that triangles with the same height and bases on the same line have areas in the ratio of those bases, and that the area of a triangle is one-half its base times its height. We call Euclid's Book 6 Prop. 2 the *Side Splitter Theorem*, on which Euclid based his triangle similarity theorems.

Theorem 1.1 (Side Splitter Theorem) *If a line parallel to one side of a triangle cuts the other two sides of the triangle, then it cuts those sides in proportion. And, conversely, a line that cuts two sides of a triangle in proportion must be parallel to the third side.*

Proof Euclid's proof depends on area, a common strategy with both Euclid and Apollonius. See Fig. 1.4 Left. We use (XYZ) to denote the area of $\triangle XYZ$. Let DE be parallel to base BC of $\triangle ABC$, where E lies on side AC and D lies on side AB. Then $\triangle DEC$ and $\triangle DEB$ have equal areas because they have equal heights and share the base DE. Now, $\triangle AED$ and $\triangle ECD$, which have the same height, have areas in the ratio of their bases AE and EC. Likewise, $\triangle AED$ and $\triangle BED$, which have the same height, have areas in the ratio of their bases, AD and DB. So $AE : EC = (AED) : (ECD) = (AED) : (BED) = AD : DB$. The same concepts are used in the proof of the converse.   □

In Book 6, we assume that when two angles of a triangle are congruent to two angles of a second triangle, then the third angles are congruent.

Proof Euclid Book 6, Prop. 4, $AA\ Triangle\ Similarity$. See Fig. 1.4 Right. We are given two triangles, ABC and DCE, with congruent corresponding angles. We place the triangles so base BC is continued in base CE, with A and D on the same side of that base. ED and BA are extended to meet at F. (That they meet is guaranteed by the Parallel Postulate.)

By Prop. 28 of Book 1, FD is parallel to AC, and BF is parallel to CD. Since opposite sides of a parallelogram are congruent (Book 1 Prop. 33), $AF = CD$

and $AC = FD$. By the Side Splitter Theorem, $AB : AF = CB : CE$. Since $AF = CD$, then

$AB : CD = CB : CE$. Likewise, since BF is parallel to CD, then
$BC : CE = FD : ED = AC : DE$. It follows that triangles ABC and DCE are similar.

□

The SSS and SAS similarity theorems are proved in the same fashion in Propositions 5 and 6 of Book 6.

We next move to Book 3 of the *Elements*, on circles. It develops propositions of importance in our examination of projective geometry. First, some terminology, followed by two important propositions.

> **Definition**
> A *tangent* is a line or segment that meets a circle in exactly one point, even if extended (Book 3 Def. 2).

These terms are not in Euclid:

A *chord* of a circle is a segment joining two points of a circle. A *diameter* is a chord that lies on the center of the circle. A *secant* is a line that meets a circle twice and extends outside the circle. An angle whose vertex is at the center of a circle is a *central angle*, and an angle whose vertex is on a circle, where each of its sides again meets the circle, is an *inscribed angle*. An *arc* is a connected part of a circle, and the *measure* of an arc is the measure of the central angle that intercepts the arc.

Euclid's Book 3 **Prop. 17** is a construction of a tangent to a circle from an outside point. **Construction A**, below, is an alternative method.

Prop. 18 (a Variation on Euclid's proof) A tangent meets a circle perpendicular to the radius drawn to the point of tangency.

Prop. 19 is the converse, whose proof makes use of (Prop. 18 (a Variation on Euclid's proof)).

Proof (Prop. 18 (a Variation on Euclid's Proof)) Let a tangent meet a circle H at point E. (The expression "circle H" denotes the circle in question with center H.) See Fig. 1.5 Left. If the tangent is not perpendicular to radius HE, then construct (Book 1 Prop. 12) the perpendicular to the tangent from point H, meeting the tangent at G. In right triangle GHE, hypotenuse EH is longer than leg HG, which means G lies inside the circle. But G is on the tangent so it cannot lie inside the circle. We conclude that the tangent at E is perpendicular to radius HE.   □

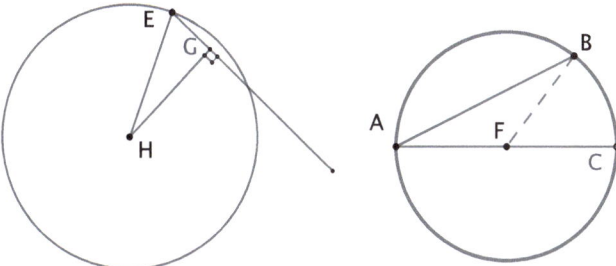

Fig. 1.5 Left: Proof of Euclid Book 3 Prop. 18. Right: Euclid Book 3 Prop. 20, Inscribed Angle Theorem. In the case pictured, a side of the angle is a diameter: $m \overset{\frown}{BC} = m\angle BFC = 2 \cdot m\angle BAC$ (The measure of exterior angle BFC is the sum of the measures of the congruent angles A and B)

Here are further theorems of importance.

Prop. 20 *Inscribed Angle Theorem.* The measure of an inscribed angle is half the measure of the intercepted arc. (See Fig. 1.5 Right.)

It follows that an inscribed angle is a right angle exactly when it intercepts a semicircle.

Construction A Let a circle have center C, and let A be an outside point. Construct a tangent from A to circle C.

Solution (Not Euclid's) Construct the circle with diameter AC. Let this new circle meet circle C in X and Y. Then lines AX and AY are tangents to circle C. (The justification is an Exercise.)

By the Inscribed Angle Theorem, we can prove the Cyclic Quadrilateral Theorem.

> **Definition**
> A quadrilateral is *cyclic* if it can be inscribed in a circle.

Theorem 1.2 (Cyclic Quadrilateral Theorem) *A quadrilateral is* cyclic *exactly when the opposite angles of the quadrilateral are supplementary.*

Prop. 32 *Angle formed by a Tangent and Chord.*
Let X and B be points on a circle. Based on Prop. 18 and Prop. 20, the angle formed by chord XB and a tangent meeting the circle at X has half the measure

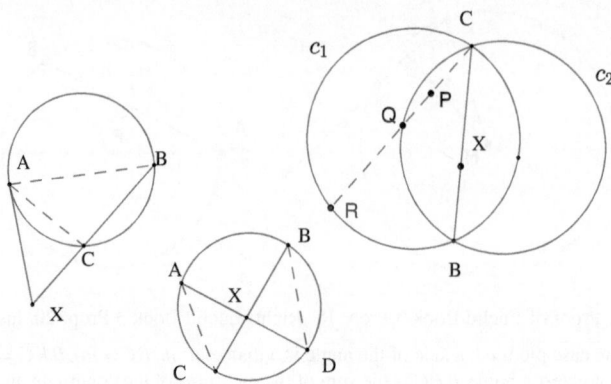

Fig. 1.6 Left and Center: Power-of-a-Point X. Right: CB is the Common Secant of circles c_1 and c_2

of the intercepted arc XB. (A hint for a proof is in the Application at the end of the chapter.)

Book 3 Propositions 35 and 36 develop the *Power-of-a-Point*, X, with respect to a given circle, the concept and name due to Jacob Steiner in the nineteenth century. The Common Secant Theorem follows [96]. See Fig. 1.6 Center for Prop. 35 and Fig. 1.6 Left for Prop. 36.

Prop. 35 If chords AD and CB meet at X, then the product $AX \cdot XD$ equals the product $CX \cdot XB$.

Prop. 36 If point X is outside the circle, and a secant on X meets the circle at C and B while a tangent from X meets the circle at A, then $BX \cdot XC = AX^2$.

Corollary to Proposition 36 (Not in Euclid): The two tangent segments drawn to a circle from an outside point are equal in length.

Modern proofs of Propositions 35 and 36 are most often based on triangle similarity. Euclid's proofs are different, as he developed the triangle similarity theorems only in Book 6.

Here is a modern proof of Euclid's Book 3, Propositions 35 and 36.

Proof Prop. 35. See Fig. 1.6 Center. Let X be inside the circle, with chords BC and AD meeting at X. The claim is that $XB \cdot XC = XA \cdot XD$. We draw AC and BD. By the Inscribed Angle Theorem, Euclid's Book 3 Prop. 20, $\angle ACB \cong \angle ADB$ and $\angle CAD \cong \angle CBD$, since the angles of each pair intercept the same arc. Therefore, $\triangle ACX \sim \triangle BDX$ by AA, so $AX : XB = CX : XD$ and the theorem follows.

In the case handled in Prop. 36, with X outside the circle, we use the same idea, with $\triangle XAC \sim \triangle XBA$, where XA is tangent to the circle. See Fig. 1.6 Left. We note, by Euclid's Book 3 Prop. 32, that $\angle XAC$ has half the measure of intercepted arc CA.   □

We can summarize in a theorem.

Theorem 1.3 (Power-of-a-Point) *Given a circle and a point X, then for any secant or chord on X meeting the circle at B and C, the value $XB \cdot XC$ depends only on X and the given circle, independent of the secant or chord BC. If a tangent from X meets the circle at A, then $XA^2 = XC \cdot XB$.*

The product $XC \cdot XB$ is called the power of X *(with respect to the circle). (The modern convention is that the power of a point X is negative for X inside the circle and positive when X is outside the circle.)*

Theorem 1.4 (Points of Equal Power Theorem) *The common secant of two intersecting circles is formed of the points whose powers with respect to the two circles are equal.*

Proof See Fig. 1.6 Right. Let the two circles, c_1 and c_2, meet at C and B. If X is on CB, then $CX \cdot BX$ is the power of X with respect to both circles. If, on the other hand, a point P is not on line CB, then line CP meets the two circles in distinct points Q and R, and the power of P with respect to one circle is $PC \cdot PQ$, which does not equal the power of P with respect to the other circle, $PC \cdot PR$.   □

Corollary 1.5 (Common Secant Theorem) *When each of three circles meets the other two, then the three pairwise common secants are concurrent (at a point called the* Radical Center of the Three Circles*). (Proof is an Exercise.)*

5   Ceva's Theorem

We have already noted that Ceva's Theorem was rediscovered several times after what seems to have been the initial discovery by Giovanni Ceva in 1678. In [21, Art. 224], Lazare Carnot proved it by two applications of Menelaus's Theorem; we will see Menelaus's Theorem in the section on Ptolemy. We give a proof here of Ceva's Theorem based on Euclid's Proposition 1 of Book 6. (We will revisit part of the theorem in the section Exercises—Homology.)

Theorem 1.6 (Ceva's Theorem) *Let concurrent lines from vertices A, B, C of a triangle meet the opposite sides, respectively, in points X, Y, Z. (Lines AX, BY, CZ are called* Cevians*.) Then $AZ \cdot BX \cdot CY = AY \cdot BZ \cdot CX$. (Note that two of the Cevians may go outside the triangle ABC.) The converse holds.*

Fig. 1.7 Left: Proof of
Ceva's Theorem. Right:
Gergonne Point

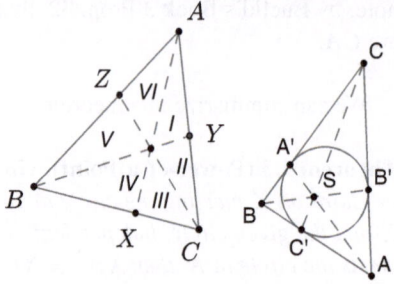

Proof See Fig. 1.7 Left. Remember that triangles with equal heights and collinear bases have areas in the ratio of the bases. So, for example, $(I) : (II) = AY : YC = (I+V+VI) : (II+III+IV)$, and, therefore, $AY : YC = (V+VI) : (III+IV)$. (Roman numerals denote the areas.) Do the same for the other two sides of $\triangle ABC$. (We use the property: If $A : B = C : D$ then $A : B = (A \pm C) : (B \pm D)$.)

$$\text{Therefore, } \frac{AZ}{BZ} \cdot \frac{BX}{CX} \cdot \frac{CY}{AY} = \frac{(I+II)}{(III+IV)} \frac{(V+VI)}{(I+II)} \frac{(III+IV)}{(V+VI)} = 1. \qquad \square$$

Example (Gergonne Point) When, as in Fig. 1.7 Right, a circle inscribed in a triangle meets the sides of the triangle at A', B', C', then Cevians AA', BB', CC' are concurrent, at a point called the *Gergonne Point of the Triangle*. The proof is an Exercise.

6   Apollonius

The ancient Greeks defined a conic section as the slice, or *section*, of a *conic surface* by a plane. A conic surface is that traced by a line, of unlimited length in both directions, resting on a point, the *vertex*, tracing a given circle in a *base plane*. The vertex cannot lie in the base plane. The part of the surface between the vertex and base plane is called a *cone*, and the slicing plane could not be parallel to the base plane. The most perfected study from the ancients was the *Conics* of Apollonius, from about 200 BCE, edited and translated into Latin in the sixteenth century, most notably by Commandinus, in 1566. The *Conics* was in eight books, of which the first seven have come down to us. In a preface, Apollonius tells us that the first four books are based on known material, most likely including lost work of Euclid.

Immediately following the definition of a conic section, Apollonius developed the concept of a *diameter* of a conic section.

> **Definition**
>
> A *chord* is a segment joining two points of a conic section. When we take all the chords of a conic that are parallel to a given chord, the set of the midpoints of those chords forms a line; such a line is called a *diameter* of the conic.

(continued)

> For a point M on a diameter, the part ML of the corresponding chord, joining M to the conic section, is the *ordinate* at M. At times, the entire chord is called an ordinate.
>
> For an ellipse or hyperbola, the midpoint of a diameter is called the *center* of the conic section.

In Fig. 1.8, taken from a 1696 edition of the Commandinus translation [5], a plane slicing the cone meets the plane of the base circle in line FG. The conic section as pictured is an ellipse on D, E, and L. BC is the diameter of the base circle that is perpendicular to FG. Then the cone is sliced by a plane on L that is parallel to the base plane, in a circle with diameter PR, as pictured. $M = PR \cap ED$. It follows that $LM \parallel FG$ and M is the midpoint of the chord on L, a chord both of the circle and of the ellipse. As this holds for all chords parallel to FG, then DE is a *diameter*. Apollonius devoted much of Book 1 to proving that all chords on the center are themselves diameters.

7   Subcontrary Circle in the *Conics* of Apollonius

Early in the *Conics*, in Book 1 Prop. 5, Apollonius presented the *subcontrary circle*, which is a particular slice of an oblique cone. (Oblique: axis from the vertex to center of base circle is not perpendicular to the base.) See Fig. 1.9, taken from Thomas Heath's [53].

Theorem 1.7 (Subcontrary Circle Theorem) *Let BC be a diameter of the circular base of an oblique cone, selected so plane ABC is perpendicular to the base plane. Slice the cone by a plane meeting AB in P and AC in P', meeting the base plane in line DE, so $\angle AP'P \cong \angle ABC$ and $DE \perp BC$. Then the slicing plane cuts the cone in a circle with diameter PP'.*

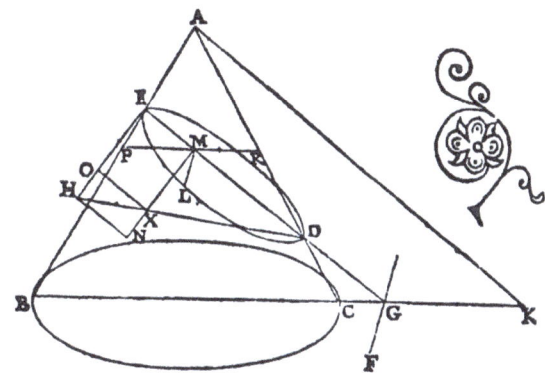

Fig. 1.8 Conic Section ELD, from the *Conics* of Apollonius, Commandinus 1696 edition (EH is the *parameter*, which we shall not refer to)

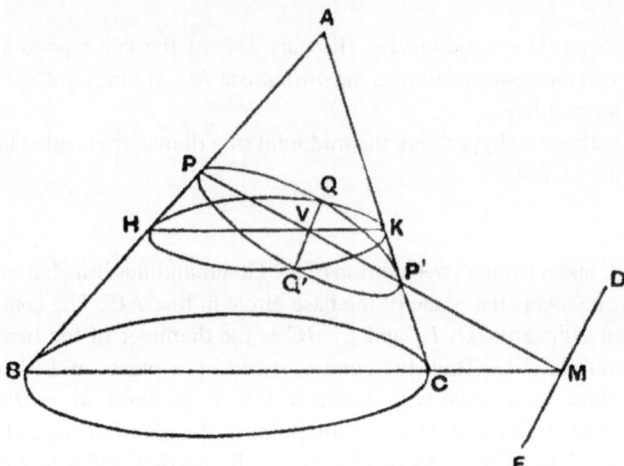

Fig. 1.9 Subcontrary Circle as developed by Apollonius, Book 1 Prop. 5. Figure is from Heath [53], p. 7

Proof Take a point V on PP' and slice the cone with a plane on V parallel to the base plane, giving a circle with diameter HK on V. On V, draw a line parallel to DE, meeting the new circle in Q and Q'. HK is the perpendicular bisector of QQ', an ordinate of the conic. Since HQK is a circle, then $HV \cdot KV = QV^2$. Because of congruent angles, $\triangle PVH \sim \triangle KVP'$ so $HV \cdot KV = PV \cdot P'V$. Since $PV \cdot P'V = QV^2$, then PQP' is a circle. (If we set $PP' = 2$, $PV = x$, $VQ = y$, then this last equation is $y^2 = x(2 - x)$, the equation of a circle.)   □

8   Ptolemy

Claudius Ptolemy worked in Alexandria, Egypt, in the first part of the second century CE. His most famous and most influential work was the *Mathematical Compilation*, usually referred to by the Arabic title conferred on it, the *Almagest*, "The Greatest." It presented the trigonometry needed for studying the heavens, explained the apparent motion of the moon and the planets, and plotted the positions of stars. With minor improvements over the centuries, it was the dominant work in planetary astronomy into the seventeenth century.

Book 1 Prop. 13 of the *Almagest* gives the earliest proof, that has come down to us, of the planar version of Menelaus's Theorem. It is the first proposition of Book 3 of the *Sphaerica* of Menelaus, from about 100 CE. That proposition is about a spherical triangle, and depends in its justification on the planar version. Desargues, among others, attributed the theorem to Ptolemy. Here are the statement and proof. (See Fig. 1.10.)

Fig. 1.10 Menelaus's
Theorem in Ptolemy Book 1
Section 13

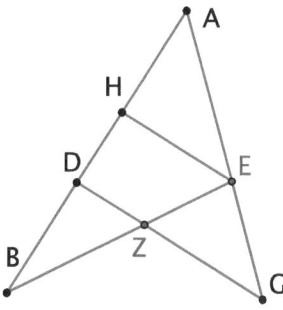

Theorem 1.8 (Meneaus's Theorem) *Given triangle EGZ, with the three sides cut by a line in A, D, and B, as pictured. Then*

$$EB \cdot GA \cdot ZD = EA \cdot GD \cdot ZB.$$

Proof Draw on E a parallel to GD, meeting AB in H. By parallels,

$$\frac{GA}{EA} = \frac{GD}{EH} \text{ and } \frac{ZD}{EH} = \frac{ZB}{EB},$$

from which the theorem follows.   □

Of further interest is Ptolemy's *Planisphere*, translated as *Flattening the Surface of the Sphere* in [89]. This is an example of a transformation between the surface of a sphere and a tangent plane to that sphere. That means we have a point-to-point pairing of the two surfaces. Ptolemy was interested in projecting the celestial sphere onto a plane. When, centuries later, that transform created a flat map of part of the surface of the earth, it was referred to as a stereographic projection. Thinking in terms of a map of the earth, one projects from the South Pole the points of the globe onto the plane tangent to the earth at the North Pole. Meridians from the North to the South Pole are represented as lines from the North Pole running off to infinity in all directions. In this sense, the single point of the South Pole of the earth is paired with a single point on the planar map, the point "at infinity," which lies on all lines of the planar map.

Ptolemy certainly knew the important property of the stereographic transform, namely, that circles on the sphere which do not lie on the South Pole are mapped to circles. Ptolemy did not offer a proof, but a proof can be based on Apollonius's exposition of *subcontrary circles* at the beginning of his *Conics*. It seems likely, but not certain, that he knew the *conformal* property of stereographic projection: lines or curves meeting at a certain angle are projected to lines or curves which meet at the same angle.

Fig. 1.11 Ptolemy's
Planisphere, based on [89]

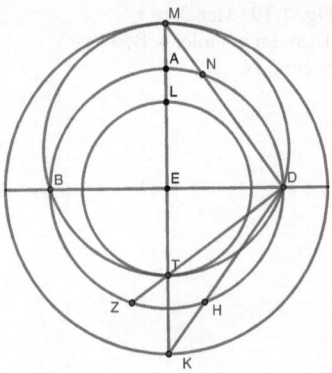

In Fig. 1.11 is a stereographic mapping discussed by Ptolemy. The sphere of the earth is projected from the South Pole onto the plane tangent at the North Pole. The circle $ADZB$ is the earth's equator. E is the earth's North Pole. We slice the sphere by a plane on the center of the sphere. That slicing plane, intended to be the ecliptic, is the plane of the solar system; it meets the equator at D and B, and its most southerly point corresponds to M and its most northerly point corresponds to T. That plane, then, is represented by circle $MDTB$. The circle with diameter MK represents the latitude line at M, suggesting the Tropic of Capricorn in the Southern Hemisphere, in which case the circle with diameter LT would be the Tropic of Cancer, in the Northern Hemisphere.

9   Optional: Circles on the Sphere Are Mapped to Circles on the Tangent Plane

Theorem 1.9 *In projecting from S, the South Pole, the sphere of the earth onto the plane tangent to the sphere at N, the North Pole, a circle on the sphere is projected to a circle on the tangent plane.*

Proof Any circle on the surface of a sphere is the intersection of some plane π with the sphere. That circle is the base of a cone with vertex S. If π is parallel to the tangent plane at N then we clearly have a circle as the projection of the circle on the sphere. For any other given circle, slice the cone by the plane α that lies on the North Pole, N, on the South Pole, S, and is perpendicular to plane π, and let plane α meet the circle in A and B. In plane α, we have the diagram of Fig. 1.12. Let line SB meet the tangent plane at point K, and let line AB meet the tangent plane at M. Now, $m(\angle MKB)$ is half of $m\widehat{NS} - m\widehat{NAB}$. So $m(\angle MKB) = (\pi - m\widehat{NAB})/2 = (m\widehat{BS})/2$. But $m \angle SAB$ is half of $m\widehat{BS}$. Since angles MKS and BAS are congruent, then by Prop. 5 of Book 1 of the *Conics* of Apollonius, about the subcontrary circle, the projection of the circle on the sphere onto the tangent

Fig. 1.12 Proof that the stereographic projection of a circle is a circle

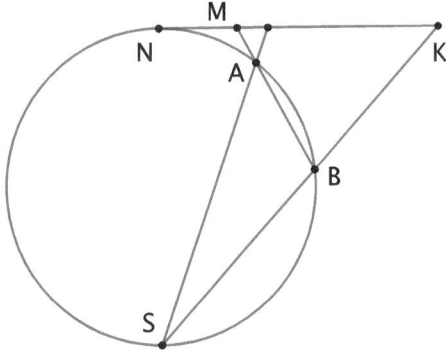

Fig. 1.13 Angles formed by chords, secants, and tangents

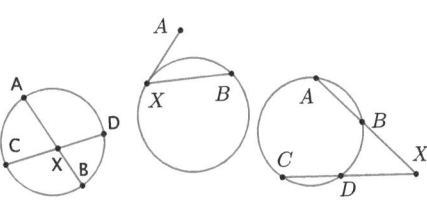

plane at N is itself a circle. (The conic surface with vertex S and on the circle with diameter AB is sliced by the tangent plane at N.)   □

10   Pappus

The last of the ancient Greek mathematicians who was an important influence on the geometry of early modern Europe is Pappus of Alexandria, who was active in the fourth century CE. His *Mathematical Collection*, edited and translated into Latin by Commandinus in 1588, is largely a summary and commentary on earlier mathematics, much of which was subsequently lost. Pappus's work has a different mathematical flavor when compared with what has come down to us from Euclid, Archimedes, and Apollonius. We put off a more detailed discussion of propositions from Pappus until we examine the early modern works he influenced.

11   Application and Exercise: Angles Formed by Chords, Secants, and Tangents

Theorem 1.10

   (i) *Let chords AB and CD meet inside a circle at X, as in Fig. 1.13 Left. Then*
   $$m\angle BXD = (m\,\overset{\frown}{BD} + m\,\overset{\frown}{AC})/2.$$

(*ii*) *The angle formed at X, on the circle, by a tangent and a chord, as in Fig. 1.13 Center, has half the measure of the intercepted arc. In Fig. 1.13,* $m\angle AXB = m\,\overset{\frown}{XB}\,/2.$

(*iii*) *When secants meet at an outside point X, the angle at X has half the difference of the measures of the intercepted arcs. In Fig. 1.13 Right,* $m\angle X = (m\,\overset{\frown}{AC} - m\,\overset{\frown}{BD})/2.$

Note that points A and B can merge, making secant XBA into tangent XA.

Hint: to *prove* (*i*) and (*iii*), draw chord BC. For (*ii*), draw a diameter on X.

12   Exercises—Greek Background

1. Prove these basic properties of a parallelogram, where a parallelogram is defined to be a quadrilateral with both pairs of opposite sides parallel.
    (*i*) (Euclid Book 1 Prop. 33) The opposite sides of a parallelogram are congruent.
    (*ii*) The diagonals of a parallelogram bisect each other.
    (*iii*) A quadrilateral is a parallelogram if both pairs of opposite sides are congruent.
    (*iv*) One of statements A and B is true and one is false.
        A. If a quadrilateral has one pair of opposite sides parallel and one diagonal bisects the other, then the quadrilateral is a parallelogram.
        B. If a quadrilateral has one pair of opposite sides congruent and one diagonal bisects the other, then the quadrilateral is a parallelogram.
    Prove the statement that is true and draw a counterexample for the statement that is false.
2. Given points A and B on a line m, and a point X not on m, construct the line on X that is parallel to m. This can be done by drawing line XA, and then constructing with vertex X an angle congruent to angle XAB placed so the two angles are corresponding angles or alternate interior angles. By Prop. 27 or 28, the new line on X will be parallel to m.
3. **Construction B.** Given points A and B and point C not on line AB, construct point D so $ABCD$ is a parallelogram. (This is an alternate way to construct a line parallel to a given line on a given point.)   Do an example of the construction.
4. Prove Euclid's Book 1 Prop. 32: An exterior angle of a triangle equals in measure the sum of the two opposite interior angles. It may help to draw on one vertex of the given triangle a line parallel to the opposite side of the triangle. Note that it follows immediately that the angle-sum of a triangles is 180 degrees.
5. Using only Euclid's Postulates of Book 1 and Propositions 1 through 28 of Book 1, show that from a point X not on a line m there is only one perpendicular line to m.

Fig. 1.14 Proof of the SSS
Triangle Congruence
Theorem, one case

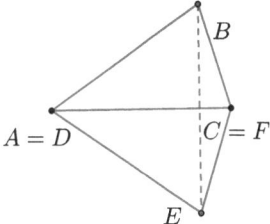

6. Finish the proof of the SSS Triangle Congruence Theorem (This is Euclid's
   Prop. 8 of Book 1, but not Euclid's proof. Use only Euclid's Postulates and
   Prop. 4, 5, and 6.)

   Given: $\triangle ABC$ and $\triangle DEF$ where $AB \cong DE$, $AC \cong DF$, and $BC \cong EF$.
   Prove: $\triangle ABC \cong \triangle DEF$.

   We will assume that A and D coincide and that C and F coincide, while B and
   E lie on opposite sides of line AC, and that line BE meets line AC between A
   and C. (See Fig. 1.14.)
7. Justify **Construction A**, i.e., prove that the constructed line AX is tangent to
   circle C.
   Hint: an angle inscribed in a semicircle must be a right angle.
   Euclid's construction is different, in Book 3 Prop. 17. Exercise: Look up
   Euclid's construction, as in [39], and explain why it works.
8. *(a)* Prove that the points of the perpendicular bisector of a segment AB are
      exactly the points equidistant from points A and B.
   (b) Prove that the perpendicular bisector of a chord of a circle lies on the center
      of the circle.
      Hint: Note the standard form of a proof that two sets are equal: first let X be
   a point of one set, say the perpendicular bisector of segment AB, and show X
   lies in the second set, in this case the set of points equidistant from A and B;
   second, let Y be a point of the second set and show it lies in the first.
9. Prove the Common Secant Theorem in the case when the three given circles
   meet each other.
   Hint: Points of Equal Power Theorem.
   It is possible that the two of the common secants are parallel. In this case, show
   that the third common secant is parallel to the other common secants; in that
   case, we can say that the common secants meet at infinity.
10. **Construction C.** (Poncelet 1813, *Cahier* 1, Problem 1, [84].)
    Given a circle C and outside points A and B, construct a circle on A and B
    which is tangent to circle C. Figure 1.15 is based on Poncelet's *Fig.* 23 of
    1813/1862, where a circle is first drawn on A and B, meeting circle C at m and
    n. Line AB meets, at P, the line on m and n. Explain why circle ABT is the
    solution.
    Hint: Apply the Common Secant Theorem.

Fig. 1.15 Based on *Fig.* 23, for Poncelet's Problem 1 of *Cahier* 1, 1813/1862

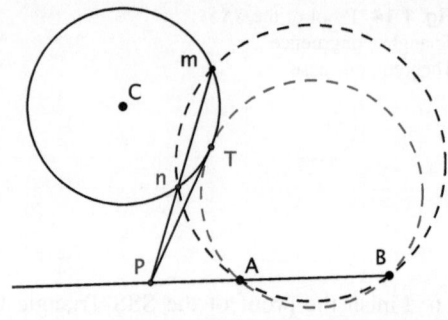

Fig. 1.16 Problem 98, from Hadamard 1906

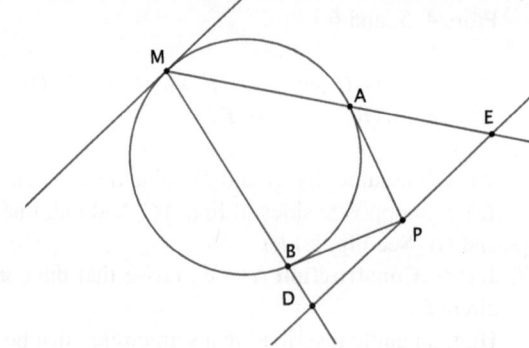

11. Given a circle on points A, B, and T, and a point P outside the circle and collinear with A and B, show that if $PA \cdot PB = PT^2$, then line PT is tangent to the circle at T. (Figure 1.15 may help.)

12. **Construction** C_1. (Poncelet 1813, *Cahier* 1, Problem 3).

    Given a line, m, and points A and B not on m, construct a circle on A and B which is tangent to line m.

    This construction can, as we shall see, be transformed to Construction C by an *inversion* transformation.

13. Jacques Hadamard 1906, page 295. Let A, B, C, D be points on a circle, with points a, b, c, d midpoints of the arcs AB, BC, CD, DA, respectively. Show that chords ac and bd are perpendicular [52].

14. Hadamard 1906, Problem 98. Given a circle and outside point P, tangents from P meet the circle at A and B, and M is any point on the circle. A line is drawn on P parallel to the tangent at M, and lines MB and MA meet this parallel line at D and E, respectively. Show that P is the midpoint of segment DE, and that length PD is the same no matter the position of M. See Fig. 1.16.

    Hint: Show $\angle AEP \cong \angle PAE$.

15. **Definition.** A *kite* is a quadrilateral with two disjoint pairs of congruent adjacent sides.

    Prove that the diagonals of a kite are perpendicular, with one bisecting the other.

Fig. 1.17 Circle of points D
such that $m \angle ADB =$
constant k or $\pi - k$

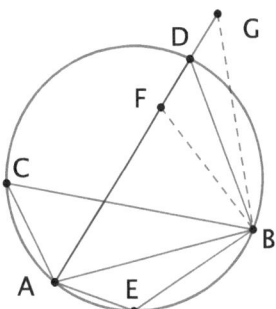

16. Given an angle with vertex X, point A on one side of the angle and B on the other side, and a length r, construct by straightedge and compass a segment WZ, W on one side of the angle and Z on the other side, so WZ has length r and is parallel to AB.

17. Based on Poncelet 1822 Art. 145. Given a quadrilateral $WXYZ$, with A on WZ, B on XY, C on YZ, D on WX, where AC, DB and WY meet at K, show

$$AW \cdot BY \cdot CZ \cdot DX = AZ \cdot BX \cdot CY \cdot DW.$$

Hint: Apply Menelaus's Theorem twice..

18. State and prove the converse of Ceva's Theorem. Note that on a line CB, a point X is uniquely determined by the ratio $CX : BX$ where the lengths are signed.

19. Gergonne Point. See Fig. 1.7 Right. Let the circle inscribed in a triangle ABC meet the sides of the triangle at a, opposite to vertex A; at b, opposite to vertex B; and at c, opposite to vertex C. Prove that the three lines Aa, Bb, Cc meet at a point. Hint: Ceva's Theorem.

   Note that the point of concurrency is called the *Gergonne Point* after Joseph Diez Gergonne (1771–1859).

20. In a triangle ABC, let lines (cevians) on B and C meet the opposite sides of the triangle at b and c, respectively, and meet each other on the median on A. Prove that $Ac \cdot Cb = Ab \cdot Bc$.

21. Prove the following. Let points A, B and C lie on a circle as pictured in Fig. 1.17. Then
    (a) if a point D lies on the circle, with $\angle ADB$ intercepting arc AB, then
       $m\angle ADB = m\angle ACB = (m\ \overset{\frown}{AB})/2$;
    (b) if a point F lies inside the circle and $\angle AFB$ intercepts arc AB, then
       $m\angle AFB > m\angle ACB$; and
    (c) if a point G lies outside the circle and $\angle AGB$ intercepts arc AB, then
       $m\angle AGB < m\angle ACB$.

22. Based on Poncelet Fig. 33, of [87, 1822]. Hexagon $ABCDEF$ is inscribed in a circle; chords BC and EF are parallel, as are chords AF and CD. Show that chords AB and DE are parallel. (Suggestion: Inscribed angles.)

23. Suppose an angle of fixed size slides along a circle, the sides of the angle tangent to the circle. Show that the locus of the vertex of the angle is also a circle.

24. Here is a simple example of a type of problem of much interest to Poncelet, problems in which a polygon is inscribed in one conic and circumscribed about another conic. Problem 16 from [86, 1817]. Suppose a quadrilateral is inscribed in one circle and circumscribed about another circle. Show that the segments joining opposite points of tangency to the inner circle meet at right angles.
    Hint. Theorem 1.10(i)

The Dilation Transformation

2

1   The *Dilation* Transformation

What do we mean by "transformation"?

> **Definition** A *transformation* of a plane is a function mapping the plane *onto* itself in a *one-to-one* fashion.
>
> (A function on a plane D is *onto* itself if every point y of D is $f(x)$ for some x in D; f is *one-to-one* if $f(x_1) \neq f(x_2)$ whenever $x_1 \neq x_2$.)
>
> We will also speak of a *transformation* from one surface to a second surface, still requiring that the mapping be one-to-one and onto.

We begin with the simplest transformation, the *dilation*. Here is the coordinate, or analytic, definition. We will soon give an alternative formulation.

> **Definition** The *dilation* with *center* at the origin and *scale factor k* maps each point (x, y) to (kx, ky), where k is real and not 0.

This definition can be adjusted so another point serves as center. Expanding or shrinking a photograph is an everyday example of a dilation.

We begin by following Jacques Hadamard's 1906 textbook *Leçons de Géométrie élémentaire* [52]. His was one of several secondary mathematics textbooks written by well known mathematicians in France—as part of school reform around the year 1900. He called the dilation a *homothetie* and figures related by a *homothetie* are *homothetic*.

C. Baltus, *Geometry by Its Transformations*, Compact Textbooks in Mathematics, https://doi.org/10.1007/978-3-031-72281-3_2

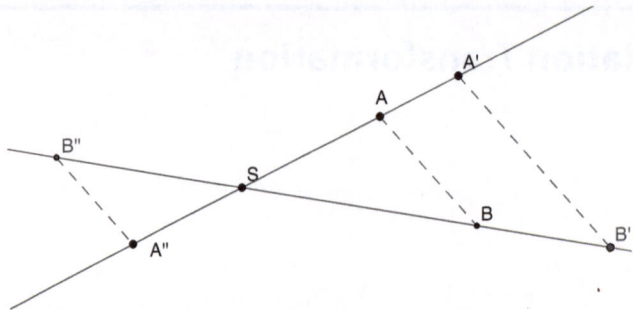

Fig. 2.1 Definition of a *dilation*, called a *homothetie* by Hadamard 1906

Definition (Hadamard) The *dilation* with *center* S and *scale factor* k, not 0, maps any point M to its *image*, M', found on line SM so the signed length of segment SM' satisfies $\frac{SM'}{SM} = k$.

A *dilation* is *direct* when the scale factor is positive, in which case M and M' are on the same side of S; it is *inverse* when k is negative, the case when M and M' are on opposite sides of S.

Two figures are *similar* when they can be placed [by a sequence of reflections] so that they are homothetic [p. 139].

(Modern Definition) Two figure are *similar* if the image of one under a dilation is congruent to the other.

In Fig. 2.1, segment $A'B'$ is the image of segment AB under a dilation with center S and positive scale factor, while $A''B''$ is the image of AB under a dilation with negative scale factor. The following theorem lists essential properties of the dilation.

Theorem 2.1 *Under a dilation with scale factor k and center S*

(i) *a segment of length l is mapped to a parallel segment of length $|k|l$;*
(ii) *since lines are mapped to parallel lines, angles are mapped to congruent angles;*
(iii) *a circle of radius r is mapped to a circle of radius $|k|r$.*

Proof (i) Since $\dfrac{SA'}{SA} = k = \dfrac{SB'}{SB}$, then by the *Side Splitter Theorem*, $AB \parallel A'B'$. And since triangles ABS and $A'B'S$ are similar, $A'B' = |k| \cdot AB$.

For (iii), all radii of a given circle under a dilation are multiplied in length by $|k|$, so the image of a circle is a circle.   □

A dilation is an element of a broader group of transformations, *collineations*, since it maps lines to lines.

Dilations are a well-known tool in geometry problem solving, the topic of I. M. Yaglom's *Geometric Transformations II* [106]. In the twentieth century, *geometric transformations* became a standard topic in secondary and college geometry courses; among these is the *dilation*. Carried out in homogeneous coordinates, the dilation is a basic transformation of computer graphics, often called *uniform scaling*. The current visibility of the topic makes it surprising that the first broad use of dilations was only in the early nineteenth century, in the work of Jean-Victor Poncelet. Most of another century would be needed to give definition to the concept.

2   Dilations in History

We recognize figures related by dilations, especially pairs of circles, in works long before a name was given to the transformation.

Our Fig. 2.2 illustrates Prop 118 from the 1588 edition of Pappus's *Mathematical Collection* [76]. The claim is that when a common tangent, KH, to two circles meets the line on the centers of those circles at a point, G, then every line on G which meets one circle must meet the other.

We find a more detailed examination of the pair of circles in the early modern era. Figure 2.3 is from François Viète's *Apollonius Gallus*, from about 1600 [103]. There are two cases of placement of the *center points* M, as Viète called them, where for secant CGM, $CM : GM = KM : LM$. On the left, M is not between the two circles, the case when the dilation is *direct*, as Hadamard called it, and the scale factor is positive. On the right, M is between the two circles, the case when the dilation is *inverse*, as Hadamard called it, and the scale factor is negative. M is a point where common tangents of the two circles meet; each circle is the image of the other under a dilation with center M.

Construction D Given two circles, construct the two *center points* of the two dilations relating the two circles. (If the circles are concentric, the common center is the one center point.)

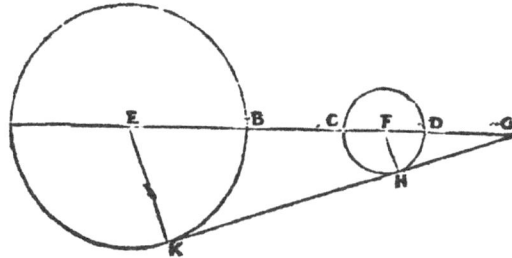

Fig. 2.2 Figure from 1588 edition, Pappus's *Mathematical Collection*, Book 7 Prop. 118

Fig. 2.3 Figure from Viete's *Apollonius Gallus*, p. 335 of [103]

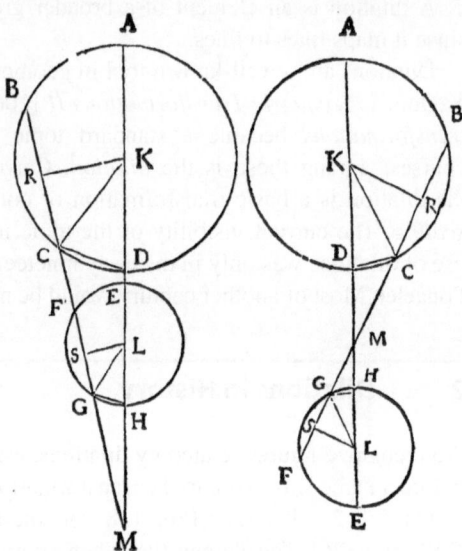

In Fig. 2.3, we construct parallel radii KC and LG. The center of the dilation is $CG \cap KL$ where K and L are centers of the circles.

Leonhard Euler, in 1748 [40], and then Poncelet, in 1820 and 1822, used the name *homologous* for points that are paired, such as G and C, and for corresponding parallel radii, LG and KC, in Fig. 2.3. The points M, corresponding to *centers of dilation*, were called *centers of similitude* by Poncelet. Euler had used the term *center of similitude* in [41], of 1777, to denote a point Γ so one figure was the image of the other by a rotation about Γ and a dilation whose center was the same point. (Euler's *center of similitude* reappeared in the twentieth century, as in [57, pp. 18–22].)

3   Gaspard Monge and Lazare Carnot

We jump in our historical narrative to the eighteenth century.

Robert Simson, in *Apollonii Pergaei Locorum Planorum Libri II*, of 1749, gave what may be the first construction in print of a common tangent to a pair of circles [95]. Like Viète, he found the point, labeled A in Simson's figure, our Fig. 2.4, that we recognize as the center of the dilation mapping one circle to the other. He constructed a tangent from A to the circle with center D, and then showed that that tangent was a common tangent.

Construction E Given two circles where neither is inside the other, construct a *common tangent* to the circles.

Fig. 2.4 Robert Simson's construction of common tangent AHK, [95, p. 6]

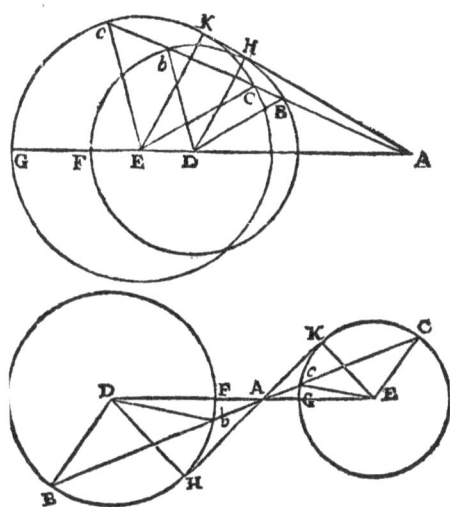

In Exercise 8, at the end of this chapter, the reader is asked to explain: If we draw a tangent to one circle from a center of a dilation relating that one circle to a second circle, then the line drawn is also tangent to the second circle.

We will concentrate on the extensive pioneering work of Jean-Victor Poncelet (1788–1867) in developing the dilation transformation early in the nineteenth century. But first we need to recognize work around the year 1800 of two French mathematicians, Gaspard Monge (1746–1818) and Lazare Carnot (1753–1823). Monge is credited with revived enthusiasm for geometry in France. From 1769 to 1784, he taught at the *École Royale du Génie de Méziers*, a military school which was then the leading institution in France for scientific studies. There he developed *descriptive geometry*. In descriptive geometry, Monge made both coordinate geometry and synthetic geometry central parts of the education of engineers. As a founder of *l'École Polytechnique*, in the 1790s, he installed a two-year sequence in geometry into the curriculum of the *École Polytechnique*. He included only a few propositions about systems of circles, but those few propositions seem to have been widely studied.

One of Monge's important results is often called Monge's Theorem.

Theorem 2.2 (Monge's Theorem)   *Suppose we are given a system of three circles of different sizes, none inside another. For each pair of circles, the two outer common tangents meet in a point, which we call a center of dilation, as in Fig. 2.5, from Monge [71] of 1799. Let those three centers of dilation be E, F, and D. Then E, F, and D are collinear.*

For each pair of circles that are exterior to each other, there are also inner common tangent lines, which meet at the second center of dilation. The theorem holds if we replace two of the points at which exterior common tangents meet, say

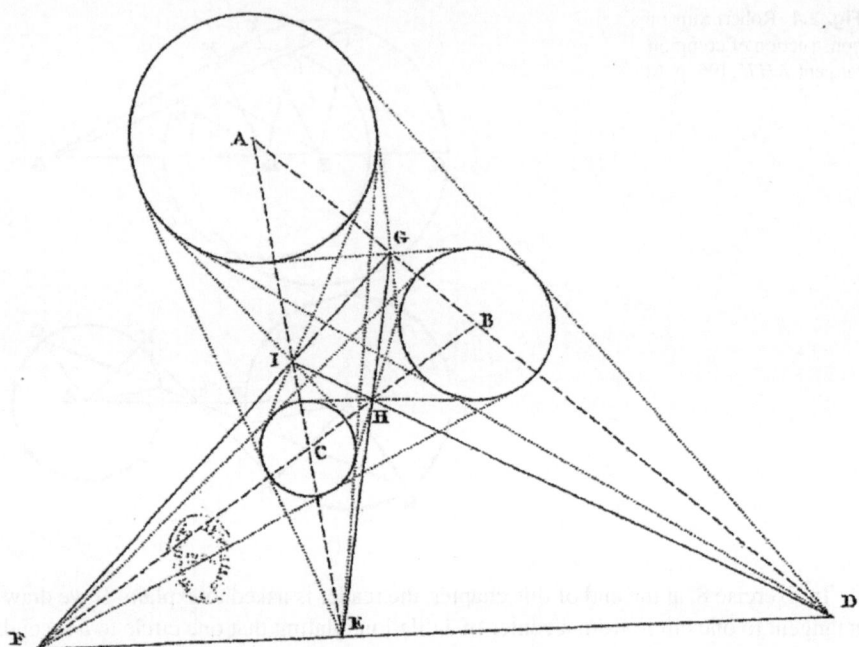

Fig. 2.5 Monge's Theorem, from *Géométrie Descriptive*, 1799

E and F, by the corresponding points at which the interior common tangents meet,
I and H. Then those three points are collinear: H, I, and D.

Monge's proof was in three dimensions, taking the spheres with centers at *A*, *B*,
and *C* which contain the three given circles, and taking a plane tangent to the three
spheres. That plane meets the plane *ABC* in a line, and that line must contain the
three centers of the dilations which, pairwise, relate the circles. We will later see a
proof in two dimensions by Poncelet.

An important theorem, known to both Girard Desargues and Philippe de la Hire
in the seventeenth century, concerns the *pole-polar* relationship. It was rediscovered
by Monge and published in 1799. We will examine the *pole-polar* relationship in a
later chapter. Here we will just say that for a point *X* outside a conic section, where
the tangents from *X* meet the conic at *Y* and *Z*, line *YZ* is the *polar* of *X* (with
respect to the conic).

Theorem 2.3 (Monge's Pole-Polar Theorem, in the Case of a Circle)   *Given a*
circle with center A and a line BC outside the circle, there is a point N inside the
circle so that for any point D on BC, the polar of D lies on N.

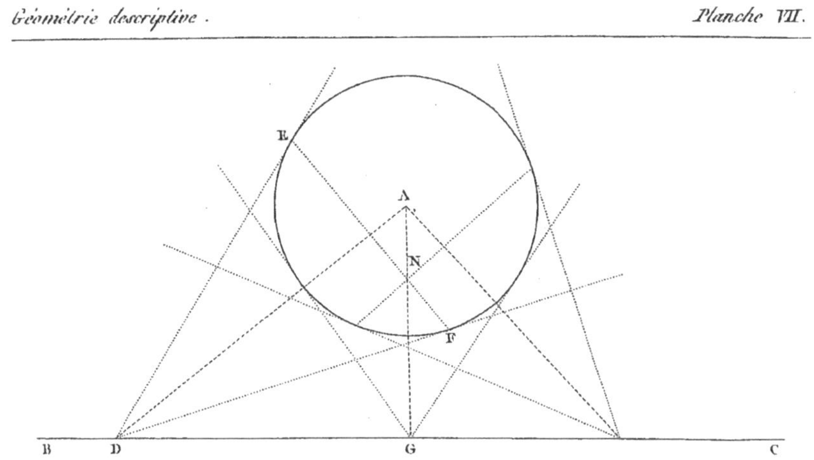

Géométrie descriptive .   *Planche VII.*

Fig. 2.6 Monge's proof, 1799, of Monge's Pole-Polar Theorem

Proof See Monge's $Fig.$ 18, our Fig. 2.6. Consider the sphere where circle A is a great circle of the sphere. Then consider the cone with vertex D on line BC and tangent to the sphere. Project onto plane ABC perpendicular to that plane. The intersection of the cone with the sphere is projected to chord EF of circle A, and EF is the polar of D.

Now consider one of the planes on line BC that is tangent to the sphere. That plane meets the sphere in a point N, whose projection onto plane ABC is also denoted N. N is also on the cone tangent to the sphere so N is on EF, the polar of D. N will be on the polar of any point on line BC.   □

Among Monge's students at Mézières was Lazare Carnot. In the French Revolution, Carnot became a member of the 12-person Committee of Public Safety in the period called the Terror. He was responsible for building the army by conscription and organizing that army to successfully defend France in 1793. Less widely known, Carnot was a mathematician. He introduced several projective concepts in his *Géométrie de Position* [21] of 1803, and even more in his 1806 *Essai sur la Théorie des Transversales* [22].

The *Common Secant Theorem* appeared in Lazare Carnot's *Géométrie de Position*: Given three circles, with centers A, B, and C, the three pairwise common secants are concurrent. Carnot's proof followed the line of Monge's proof of Monge's Theorem. One considers the spheres whose equators are the three given circles. A projection perpendicular to plane ABC maps the pairwise intersections of the spheres onto the common secants, which necessarily meet in a single point.

In *Géométrie de Position*, Carnot introduced *signed lengths*. On a number line, AB would denote the coordinate of B minus the coordinate of A. So for collinear points A, B, C, $AC = AB + BC$ no matter the order of the points. With signed lengths, the *power-of-a-point* X with respect to a circle is $XA \cdot XB$ for any line on X that meets the circle in points A and B.

Also in *Géométrie de Position*, Art. 334, Carnot gave a coordinate solution of the Problem of Apollonius: to find a circle tangent to three given circles. We will soon examine Poncelet's solution to the same problem.

Carnot's 1806 *Essai sur la Théorie des Transversales* was an influential work that demonstrated many of the theorems of projective geometry. Although the methods were not projective, it highlighted problems amenable to projective solutions, problems that would reappear in Poncelet's work.

4   Dilations in Poncelet

Now we will follow the work of Jean-Victor Poncelet. He is best known for introducing projective geometry, in *Traité des propriétés projectives des figures*, published in 1822. But he was also the first to fully develop the dilation, which he called a *similitude*, in his first *Cahier* of 1813, which was only published in 1862, and in briefer form in his *Essai* of 1820 and his *Traité* of 1822 . He called two figures related by a dilation *semblable et semblablement placées, similar and similiarly placed*, abbreviated *s. et s. p.*

Poncelet entered the *École Polytechnique* in 1807 and he graduated in 1810 as an officer in the Engineering Corps. He went with Napoleon's army into Russia in 1812, and he was captured in the wintry retreat from Moscow late in the year. A long march brought him to a military prison at Saratoff. He recovered his health in the spring of 1813 and began a series of seven mathematical *Notebooks*, or *Cahiers*. He had no books to work from, but clearly had deeply imbibed the geometry of the *École Polytechnique*. The first Notebook included a study of pairs of circles that we think of as related by a dilation [84]. In the other notebooks, he worked out in preliminary form the projective geometry that would fill his book of 1822. The dilation is a special type of projective transformation, which we will later see.

In 1809, as an student, Poncelet wrote a paper that appeared in the 1813 *Correspondance sur l'École polytechnique* [82]. The paper was a solution of the Problem of Apollonius. The solution involves pairs of circles related by dilations. Properties employed in the proof were attributed to a short article of J. N. P. Hachette (1769–1834) [51, p. 20], from the May 1807 issue of the *Correspondance sur l'École Imperiale Polytechnique*.

The mathematical background called on in the 1809 proof was only systematically developed in Poncelet's *Cahier* 1 of 1813, which Poncelet titled *Lemmes de géométrie synthetique: sur les systèmes des circles situés dans une mème plan*. We will look at propositions in Poncelet's first notebook, of 1813, and then follow Poncelet's application of those proposition in his 1809 solution to the Problem of Apollonius.

Theorem 2.4 (Poncelet 1813, Cahier 1 Prop. 1. See Fig. 2.7 Left) *Let C_1 and C_2 be centers of circles with radii, respectively, r_1 and r_2, with parallel radii $C_1 P_1$ and*

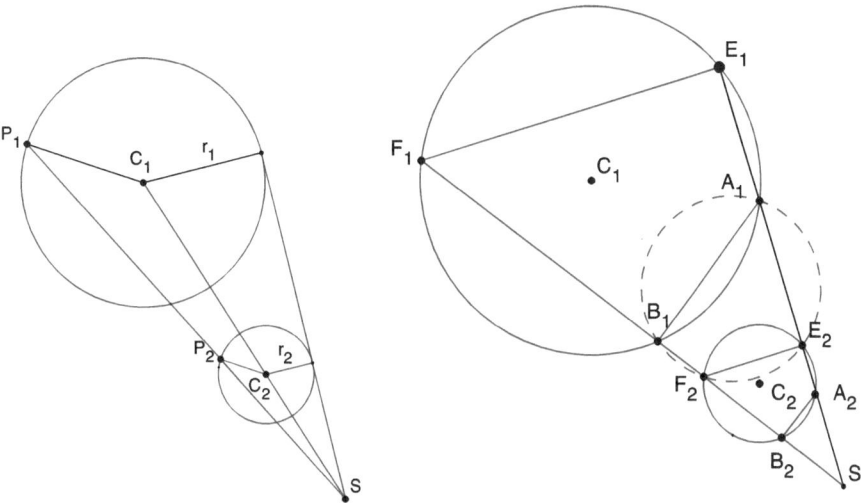

Fig. 2.7 Poncelet 1813. Left: Prop. 1. Right: Prop. 3

$C_2 P_2$, *as pictured. Let line $P_1 P_2$ meet line $C_1 C_2$ at S. Then by Euclid's Book 6 Prop. 2, the Side Splitter Theorem, and triangle similarity,*

$$\frac{r_1}{r_2} = \frac{C_1 S}{C_2 S} = \frac{P_1 S}{P_2 S},$$

and we have the same result with the same point S for all other choices of parallel radii $C_1 P_1$ and $C_2 P_2$.

Proposition 1 was followed by the *Lemme Général*, extending the result beyond circles, to triangles.

Theorem 2.5 (General Lemma) *If corresponding triangles ABC and abc have corresponding sides parallel, then the lines on corresponding vertices, Aa, Bb, and Cc meet at a single point S. See Fig. 2.8.*

Proof Poncelet observed that the triangles are similar, by AA. Let Aa meet Bb at O. By triangle similarity and the *Side Splitter Theorem*,
    $BO : bO = AO : aO = AB : ab = CB : cb$. For the same reason, if Bb and Cc meet at a point O^*, then $BO^* : bO^* = CO^* : cO^* = CB : cb$. This means O and O^* are the same. In our language of dilations, we add an observation that Poncelet did not: the scale factor for this dilation, with center O and mapping A to a, is Oa/OA, and all segments and their images are in this ratio.   □

The second proposition of Poncelet's *Cahier 1* is *Monge's Theorem*, with a two-dimensional proof.

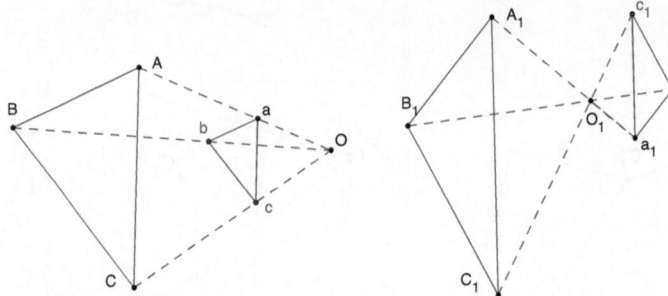

Fig. 2.8 Poncelet *Cahier* 1, General Lemma, based on Figs. 2.2 and 2.3, 1813/1862

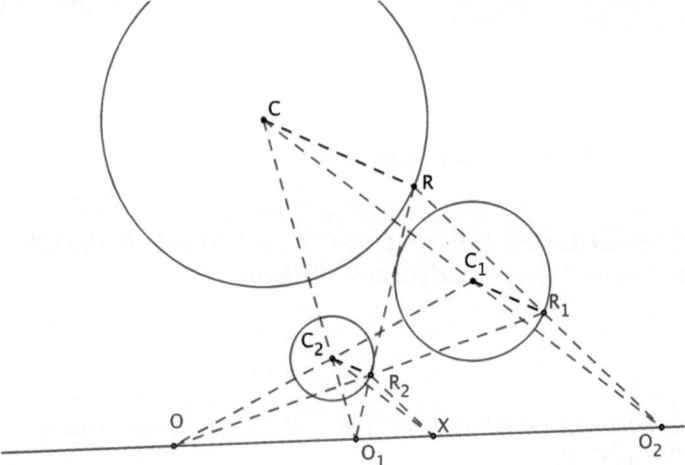

Fig. 2.9 Poncelet's proof of Prop. 2, 1813, Monge's Theorem

Theorem 2.6 (Poncelet 1813, Cahier 1 Prop. 2.) *Given three circles with centers* C, C_1, C_2, *with corresponding parallel radii* CR, C_1R_1, C_2R_2, *and with respective meeting points*

$$O = C_1C_2 \cap R_1R_2, \quad O_1 = CC_2 \cap RR_2, \quad O_2 = CC_1 \cap RR_1. \text{ Then } O, O_1, O_2$$
are collinear.

Proof See Fig. 2.9. At C_2 we draw a line parallel to CC_1 and at R_2 a line parallel to RR_1, where these two new lines meet at X. By the General Lemma following Prop. 1, applied to triangles $R_1O_2C_1$ and R_2XC_2, line O_2X lies on O. Likewise, triangles RO_2C and R_2XC_2 have corresponding sides parallel, so O_1 lies on line XO_2. So O, O_1, and O_2 are collinear.   □

Poncelet's Proposition 3 involves *cyclic quadrilaterals*. As developed in Theorem 2 of Chapter 1, a quadrilateral is *cyclic* if it can be inscribed in a circle, which holds exactly when a pair of opposite angles are supplementary.

Theorem 2.7 (Poncelet 1813, Cahier 1 Prop. 3 and Scholie) Proved by Hachette in 1807 [51]. See Fig. 2.7 Right. *Let C_1 and C_2 be centers of circles, with center of similitude S, and inscribed quadrilaterals $A_1B_1F_1E_1$ and $A_2B_2F_2E_2$ corresponding under a dilation with center S. Then*

(i) Quadrilaterals $A_1B_1F_2E_2$ and $A_2B_2F_1E_1$ are cyclic quadrilaterals, and
(ii) $SA_1 \cdot SE_2 = SB_1 \cdot SF_2$, where these products are the same no matter how we select inscribed quadrilateral $A_1B_1F_1E_1$.

Proof Let r_1 and r_2 be the radii of the circles. We note that chords F_1E_1 and F_2E_2 are parallel, giving congruent corresponding angles at F_1 and F_2. Following pairs of supplementary angles proves (i). For (ii), note that

$$\frac{SE_1}{SE_2} = \frac{r_1}{r_2} = \frac{SF_1}{SF_2},$$

and $SE_1 \cdot SA_1 = SF_1 \cdot SB_1$ by the power-of-point S with respect to circle C_1. So

$$\left(SE_2 \cdot \frac{r_1}{r_2}\right)SA_1 = \left(SF_2 \cdot \frac{r_1}{r_2}\right)SB_1.$$

Then simplify. $\qquad\qquad\qquad\qquad\qquad\qquad\qquad\qquad\qquad\qquad\qquad\qquad\square$

In Fig. 2.7 Right, fix the two circles and center of dilation S, and let B_1 slide along circle C_1 until it coalesces with A_1, in the process letting points B_2, F_2, and F_1 also slide along their circles. Then the circle on A_1, B_1, F_2, and E_2 becomes tangent to the two given circles at points A_1 and E_2, with S, A_1, and E_2 collinear. We then arrive at the following corollary.

Theorem 2.8 (Poncelet 1813, Corollary to Prop. 3) *Let c_1 and c_2 be circles, neither inside the other, which are paired by a dilation with center S. Then when a third circle is tangent to circle c_1 at T_1 and tangent to circle c_2 at T_2 we have S, T_1, and T_2 collinear.*

Poncelet's **Problem 2** at the end of *Cahier* 1 is the Problem of Apollonius. We, however, present the solution of that problem as in Poncelet's paper of 1809.

Construction F *Problem of Apollonius*: Construct a circle tangent to three given circles.

Fig. 2.10 Poncelet's 1809
proof of the Problem of
Apollonius

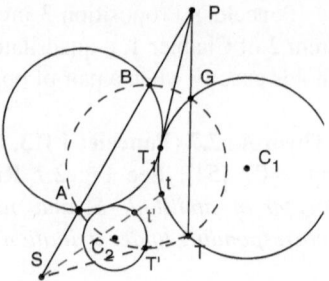

We will see that this construction is equivalent to: Construct a circle tangent to two given circles and an outside point.

Modern solutions of the Problem of Apollonius go back to François Viète (1540–1603), around 1600, in *Apollonius Gallus* [103, pp. 334–336]. Poncelet's 1809 solution was in synthetic geometry, as was Viète's. (Viète had begun with what are essentially Propositions 1 and the Proposition 3 Scolie of Poncelet's *Cahier* 1.) Poncelet would later return to the problem—a sort of test case in geometry—several times.

First, as had Viète, the radii of the three given circles are reduced by the least of the three radii, so one of the circles becomes a point, A. When we have found the circle on A that is tangent to the two other circles, with centers C_1 and C_2, as in Fig. 2.10, the radius of that found circle will be decreased by the amount of the original shrinking.

Lemma 2.1 *We are given circles C_1 and C_2 with S the center of dilation relating those two circles, and point A. We take, as pictured in Fig. 2.10, T on circle C_1 and T' on circle C_2, collinear with S. Then there is a point B on line SA, independent of the choice of T and T', that lies on all of the circles ATT'.*

Proof By Poncelet's Prop. 3, the product $ST \cdot ST'$ is the same for all choices of T and T'. (The product is $St \cdot St'$ in Fig. 2.10.) But $ST \cdot ST'$ is the power of S with respect to circle ATT', and must equal $SA \cdot SB$ when B is the second point at which line SA meets circle ATT'. So B is determined.   □

Now let G be the point, in addition to T, at which circle ATT' meets circle C_1. Let P be the intersection of lines TG and SA. Let T_1 be the point at which a tangent from P meets circle C_1. (There are two such points.) By the power of point P with respect to circles C_1, ABT_1, and ATT',

$$PA \cdot PB = PG \cdot PT = PT_1^2.$$

Then $PA \cdot PB = PT_1^2$ is only possible for the circle on A, B, and T_1 when line PT_1 is a tangent to that circle. Thus, that circle and circle C_1 are tangent at T_1. By the Corollary to Prop. 3, above, circle ABT_1 is tangent, also, to circle C_2 at a point of C_2 lying on line ST_1.   □

5   Application 1 of the Dilation: The Square in a Triangle

Solutions to a straightedge-compass problem illustrate the advantage provided by the dilation transformation. We can describe the strategy thus: First, form a figure that partially satisfies the requirements of the figure to be found, then apply a dilation to arrive at the figure desired.

Construction G   Inscribe a square, $MNPO$, in a given triangle ABC so P and O lie on side BC.

There is a 1760 solution by Robert Simpson [94, p. 210]. We turn to Carnot's 1803 solution [21, $Fig.$ 70], which does not use a dilation. See Fig. 2.11. A square $mnpo$ is constructed. Then a line ab is constructed on vertex m so $\angle amn$ is congruent to the given $\angle B$. Presumably we, likewise, construct an angle at n congruent to the given $\angle C$. This permits us to finish the construction, as pictured, of square $mnpo$ inscribed in triangle abc, similar to the desired result. Then, as Carnot tells us, we have the proportion $MO : mo = BC : bc$, which lets us find the ratio $MO : mo$. Since all corresponding sides are in this ratio, we can carry out the desired construction.

The simpler construction, now well known, is to pick a point m_1 on side AB, then complete the construction of square $m_1 n_1 p_1 o_1$, with o_1 and p_1 on side BC. Now perform a dilation with center B, mapping n_1 to a point, N, on side AC. Using that N, complete the construction of image square $NPOM$, as pictured in Fig. 2.11 Right. This construction can be found, for example, in Yaglom's [106].

Carnot's elaborate construction in 1803 suggests that dilation was not then a common tool in problem solving.

(There is an 1811 solution [43] by L. A. S. Feriot, simpler than that of Carnot.)

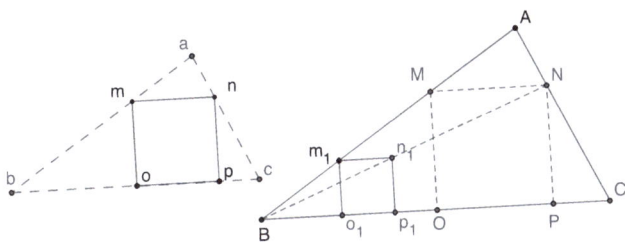

Fig. 2.11 Left: Based on Carnot's Figures 70 and 71, 1803. Right: Based on Yaglom's Figure 73, p. 94, of [106]

Fig. 2.12 Pappus
Proposition 130 of Book 7

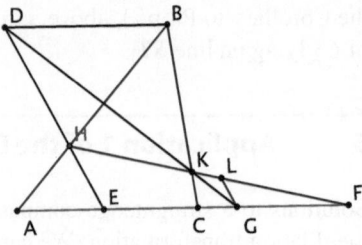

---

6   Application 2 of the Dilation: Pappus Book 7 Proposition 130

Here is an application of dilations to a proof problem, where the scale factor is negative. This is Proposition 130 of Pappus's Book 7 [76], with the diagram based on Commandinus's 1588 Latin edition. It is not Pappus's proof. This is an example of a type of problem common in the work of Pappus, proving that three certain points are collinear, a type of problem not found in Euclid's *Elements*.

> Let $ABCDEFGH$ be a figure with AF parallel to DB, and $AE : EF = CG : GF$ [for collinear points A, E, C, G, F.] [Let H be $DE \cap AB$, $K = DG \cap BC$.] Then I say that a line lies on H, K, and F.

See Fig. 2.12. Draw LG on G parallel to DE, so L is collinear with H and K but we do not assume F is collinear with H and K. Join C and L with a segment (not shown in Fig. 2.12). First, apply the dilation with center K mapping B to C. Since $DB \parallel CG$, then D is mapped to G. Likewise, H is mapped to L. This means that diagonal BH is mapped to CL, and so $BH \parallel CL$.

Next, consider the dilation with center F mapping G to E. Since $AE : EF = CG : GF$, then C is mapped to A. As lines are mapped to parallel lines, then the image of L lies on DE. But $CL \parallel BH$, so the image of L lies on BA. Thus, L is mapped to H, so H, L, and F are collinear. Therefore, H, K, and F are collinear.

□

---

7   Application 3 of the Dilation: The 9-Point Circle, with Exercises

The 9-Point Circle was developed by Leonhard Euler [42] using coordinate geometry. Lazare Carnot gave a proof in synthetic geometry in [21, Art. 131] of 1803. We present a development by the dilation transformation, as found in Coxeter's [30] *Introduction to Geometry*, based on dilations. It closely follows a proof by F-J. Servois [93] in 1804, although in Servois' work the dilations are not explicit. The theorem involves three centers of a triangle: the orthocenter, centroid, and circumcenter. Part of the claim is that those centers are collinear. The 9-Point

Fig. 2.13 S is the Centroid
of $\triangle ABC$. AY and BX are
medians

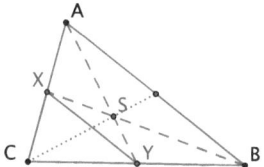

Circle is the image, by a dilation, of the circle circumscribed to a triangle, and the
nine points are related to the three centers mentioned.

We begin with a couple preliminary lemmas.

The Centroid

A *median* of triangle is a line joining a vertex of the triangle to the midpoint of
the opposite side. Archimedes, in the third century BCE, recognized that the three
medians of a triangle all meet at one point, and that that point is the *centroid*, or
center of mass, of the triangle [7, On the Equilibrium of Planes I].

Lemma 2.2 *The medians of a triangle are concurrent, at the* centroid *of the
triangle; they meet at a point on each median whose distance from the vertex is
twice the distance from the opposite midpoint.*

Proof See Fig. 2.13. As pictured, X and Y are midpoints of sides AC and
BC, respectively, of triangle ABC. We draw segments AY, BX, and XY, with
$AY \cap BX = S$. Since XY cuts two sides of the triangle in proportion, it is
parallel to the third side, AB. Since $AB \parallel XY$, then we have congruent alternate
interior angles and congruent corresponding angles, making $\triangle XYS \sim \triangle BAS$ and
$\triangle XYC \sim \triangle ABC$ with, in each case, corresponding sides in the ratio of 2 : 1. The
ratio 2 : 1 holds since the X and Y are midpoints of sides of the triangle. Thus S
divides medians AY and BX in ratio 2:1. If we repeat the argument using a different
pair of medians, those medians will also be divided in the ratio 2:1, forcing the three
medians to be concurrent.   □

The Circumcenter

Lemma 2.3 *Any triangle ABC is circumscribed by a circle whose center, the*
circumcenter, *is the point at which the three perpendicular bisectors of the triangle
sides meet. In particular, the perpendicular bisectors of the three sides of a
triangle are concurrent, at the* circumcenter *of the triangle. (The construction of
the circumcenter is treated in Exercise 6 at the end of this chapter.)*

The Orthocenter

Lemma 2.4 *The* altitudes *(lines perpendicular to the opposite sides from the three
vertices) are concurrent at a point, O, called the* orthocenter.

Fig. 2.14 9-Point Circle

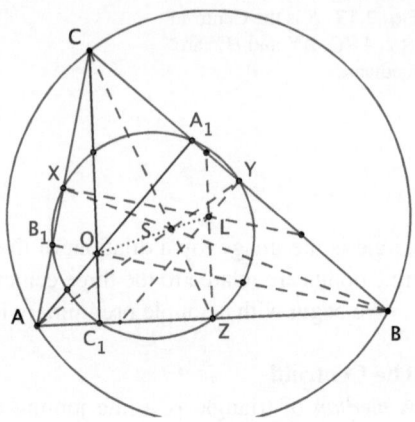

Proof Given $\triangle ABC$, let X, Y, and Z be midpoints of the sides, as pictured in Fig. 2.14, and S the centroid. We have drawn the perpendicular bisectors of two sides, meeting at the circumcenter, L.

We carry out a dilation with scale factor -0.5 and center S. Since S divides the medians in ratio 2:1, then the vertices are mapped to the midpoints of the opposite sides, and the sides of the triangle are mapped to the segments joining midpoints of the sides, creating $\triangle XYZ$. The images of the midpoints are themselves midpoints of the sides of $\triangle XYZ$.

Problem 1 What are the images of the altitudes of $\triangle ABC$? What are these images with respect to $\triangle XYZ$? What are these images with respect to $\triangle ABC$? Why must these images be concurrent? Recall that perpendicular lines will be mapped to perpendicular lines.

Solution Since $\triangle ABC$ is mapped to $\triangle XYZ$, then the altitudes of $\triangle ABC$ are mapped to the altitudes of $\triangle XYZ$. The altitudes of $\triangle XYZ$ are perpendicular bisectors of the sides of $\triangle ABC$. So they meet at the circumcenter of $\triangle ABC$.

By this exercise, we see that the altitudes of $\triangle ABC$ are mapped to concurrent lines. Since a dilation is a one-to-one mapping, we can conclude that altitudes of $\triangle ABC$ are concurrent, at the *orthcenter*.   □

Problem 2 Now, explain why the circumcenter, the orthocenter, and the centroid are collinear.

Solution Since a point and its image under a dilation are collinear with the center of the dilation, then the circumcenter, the orthocenter, and the centroid are collinear.

                                                                                    □

Definition The image of the circle on A, B, and C under the dilation whose center is the centroid of $\triangle ABC$ and scale factor is -0.5 is the *9-Point Circle* of triangle ABC.

What is the center of the *9-Point Circle*?

Solution The 9-point circle of $\triangle ABC$ must lie on the images of the three vertices, i.e., it lies on X, Y, and Z. The center of the 9-point circle must be L' since L is the center of the circle on A, B, C.

Problem 3 Describe the location of L' on line OL. Then explain why the *feet*, A_1, B_1, and C_1, of the altitudes of $\triangle ABC$ lie on the 9-Point Circle.

Solution L' is on line OL since O, S, and L are collinear. Then the center of the dilation, S, is collinear with L and L', where length OL' is half of OL. LS is half of OS, so L' is the midpoint of OL.

We now look at trapezoid OC_1ZL, where C_1 is the foot of the altitude from C. There are right angles at C_1 and Z, and L' is the midpoint of side OL. Thus, we have $L'C_1 = L'Z$. This means that C_1 is on the circle with center L' and lying on Z. So C_1 (with the other feet of the altitudes) is on the 9-Point Circle. □

To explain the remaining triple of points on the 9-Point Circle, we will apply a different dilation to the circle on A, B, and C. This time the center is O, the orthocenter, and the scale factor is $+0.5$. The center of the image circle is L', as found above, the midpoint of segment OL. From our work above, we see that this new image circle has the same center as the 9-Point Circle, and its radius is half the radius of the circumscribed circle. So the new image circle is the 9-Point Circle. And that image circle is on the images of A, B, and C. In summary:

Theorem 2.9 *The 9-Point Circle of $\triangle ABC$, defined to lie on the midpoints of the sides of $\triangle ABC$, includes the feet of the three altitudes and the midpoints of the segments joining the orthocenter to the three vertices of $\triangle ABC$.*

8   Application 4 of the Dilation: π and the Area of a Circle and Liu Hui

We will consider a *curve*, f, of the plane to be a continuous function, or the graph of such a function, mapping a finite interval $[a, b]$ onto a set of points in $\mathbb{R}^2$. It is *rectifiable* if there is a finite bound that exceeds the lengths of all polygonal paths formed by joining in order a finite collection of points on the curve. In other words, a curve is rectifiable if there exist a bound M so for any (ordered) partition $\{a_0 = a, a_1, a_2, \ldots, a_n = b\}$ of $[a, b]$

$$\sum_{j=1}^{n} |f(a_j) - f(a_{j-1})| \leq M.$$

The *length* of the curve is the *least upper bound* of these sums. Because a dilation with scale factor k multiplies all segment lengths by $|k|$, we have this theorem:

Theorem 2.10 *A dilation of scale factor k multiplies the length of any rectifiable curve by $|k|$.*

Without proof, we note that any circle is rectifiable, and the length of the curve is called the *circumference* of the circle. As any two circles are related by a dilation, and a dilation multiplies both the diameter of a circle and its circumference by the same factor, the first part of the following theorem holds. Proof of the second part follows.

Theorem 2.11

(i) *The ratio of the circumference of a circle and its diameter is the same for all circles. This ratio is denoted π.*

(ii) *The area of a circle is one-half the product of its radius and its circumference..*

As noted in the Introduction, Archimedes proved that the ratio of the circumference of a circle to its diameter is between $3\frac{10}{71}$ and $3\frac{1}{7}$, and that the area of a circle and the square of its radius are in the same ratio [7, Measurement of a Circle].

We present the exposition of Liu Hui relating the area of a circle to its circumference and his approximation of π, comparable to Archimedes' work. Liu Hui's proof is from his commentary on the book *Jiuzhang Suanshu*, or *The Nine Chapters on the Mathematical Art*. (Jiu = nine, zhang = chapters, suan = computation or mathematics, shu = method or art or prescription.)

The *Jiuzhang Suanshu* was compiled in the Han Period in China (206 BCE–220 CE), with the current version probably from the first century CE. It is a collection of 246 problems with solutions, arranged in nine groups. Its authors are unknown. Several known commentators supplied explanations and corrections; Liu Hui gave the first important commentary, from 263 CE.

The little we know about Liu Hui is from his introduction to his commentary on *The Nine Chapters*. He must have lived in the Kingdom of Wei, one of the three kingdoms established in China after the collapse of the Han Empire in 220 CE.

Several ancient versions of the *Jiuzhang Suanshu* and commentaries were collated by Qian Baocong and published in 1964. This book is the source for the 1999 [74], the basis for this section.

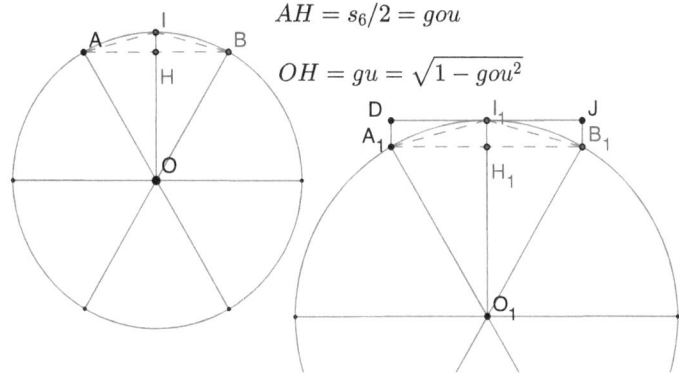

Fig. 2.15 Left: The inscribed 6-gon and 12-gon. Right: Part of a circumscribed polygon.

Here is **Problem 31** of Chapter 1 of the *Jiuzhang Suanshu*.

Now given a circular field, the circumference is 30 bu and the diameter 10 bu.
Tell: What is the area?
Answer 75 [square] bu.

Liu Hui began his commentary by considering a sequence of inscribed regular polygons, of 6 sides, then 12 sides, then 24, and so on. For simplicity we set the radius of the circle at 1. He used a base-ten numeration system, where the basic unit was the *chi*, about 32 cm, and then 1 *chi* = 10 *cun* and so on, with a new unit for each of the distinct decimal places; we will translate into our decimal system. The inscribed hexagon is formed of six equilateral triangles whose sides are 1 unit, giving perimeter of 6 and area of $6 \cdot \dfrac{\sqrt{3}}{4} = \dfrac{3}{2}\sqrt{3}$. Liu Hui immediately pointed out that the circumference of a circle of diameter 2 cannot be 6, since 6 is the perimeter of the inscribed hexagon.

We point out important theorems employed in Liu Hui's computations.

First, the Chinese had for centuries known the Pythagorean Theorem, which they called the *Gougu* Theorem: $gu^2 + gou^2 = xian^2$ where gu is the longer leg of a right triangle, *gou*, or "hook," the shorter leg, and *xian*, or "bowstring," the hypotenuse.

Second, the area of a kite is half the product of the two diagonals. (This follows from Exercise 15 of Chap. 1.)

Third, the Chinese were long able to compute square roots to any desired accuracy, a skill they shared with ancient Babylonian and Greek mathematicians.

Now consider the inscribed 12-gon, superimposed on the hexagon, as in Fig. 2.15 Left. We let s_6, which equals 1, be the side of the 6-gon, $p_6 = 6$ the perimeter of the 6-gon, and $A_6 = \dfrac{3}{2}\sqrt{3}$ the area of the 6-gon. (Liu Hui gave $\sqrt{3}/2$ as 0.8660254.)

Let us now find s_{12}, p_{12}, and A_{12}, with respect to the inscribed 12-gon. As $AO = 1$ and $AH = 1/2$, then $OH = \sqrt{3}/2 = gu$ and $HI = 1 - \sqrt{3}/2 = 1 - gu$. Then, where gu and gou are with respect to triangle AHO,

$$s_{12}^2 = AI^2 = gou^2 + (1 - gu)^2 = gou^2 + gu^2 + 1 - 2gu = 2 - 2gu$$

$$= 2 - 2\sqrt{1 - s_6^2/4} = 2 - \sqrt{4 - s_6^2}.$$

In this way we have the recursive formula

$$s_{2n}^2 = 2 - \sqrt{4 - s_n^2}.$$

Liu Hui computed $s_{96} = 0.065438$, correct to the sixth decimal place. (Suggestion to the reader: By calculator, show that by finding the sequence of side lengths, you arrive at the same value for s_{96}.)

How about the area of the inscribed figures? In Fig. 2.15 Left, $AIBO$ is a kite, whose area is $1 \cdot s_6/2$. A_{12} is 6 times the area of the kite, which equals one-half the perimeter of the 6-gon, p_6. In the same way, A_{192} is one-half p_{96}. Liu Hui computed A_{96} to be $313\frac{584}{625}sq.\,cun = 3.139344$ and A_{192} to be $314\frac{64}{625}sq.\,cun = 3.141024$.

Treating the area of the circle as the limiting value of the area A_{2n} as $n \to \infty$, we see that the area is one-half the circumference, when the radius is 1, and $0.5 \cdot r \cdot c$ when the circle has radius r and circumference c.

The area of an inscribed polygon is less than the area of the given circle. The inscribed polygon Liu Hui considered was the regular 192-gon. How do we find a circumscribed polygon that will be close in area to the circle? See Fig. 2.15 Right. Adding the rectangles A_1DJB_1 to the sides of the inscribed hexagon produces a polygon circumscribed to the circle. The area of all the added rectangles is twice the difference of the areas A_{12} and A_6, so the area of the entire circumscribed polygon is $A_6 + 2(A_{12} - A_6) = 2A_{12} - A_6$. Applying the same reasoning to the 96-gon and the 192-gon, we have a circumscribed polygon of area $2A_{192} - A_{96} = 314\frac{169}{625}sq.\,cun = 3.142704$. Thus, Liu Hui felt justified in using the approximate value 3.14 for π. He went further, by a different method, to arrive at the approximation 3.1416.

9   Exercises—Dilations

1. Under the dilation with center $S = (2, -3)$ and scale factor 4, what is the image of
   (i) line $y = 3x - 2$,
   (ii) circle $y^2 = x(4 - x)$.

2. For circle c_1, with equation $x^2 + y^2 = 4$, and circle c_2, with equation $(x - 10)^2 + y^2 = 49$, find the two centers of dilation that map one circle to the other. Then find the equation of one common tangent not between the two circles.

3. Let m and m' be parallel lines, and let n and n' be parallel lines, where m and n meet in point X while m' and n' meet in Y. Find the center and scale factor of the dilation that maps m to m' and maps n to n'.

4. **Construction H** Given an angle $\angle AXB$ and a point P inside the angle, construct by straightedge and compass a circle on P which is tangent to the two sides of the angle. Suggestion: First draw some circle tangent to the two sides of the angle, then apply a dilation so the image has the properties sought. There will be two such circles.

5. Exercise 5 asks for an alternate construction to **Construction C** of the previous chapter.
   Construction C_1 Given a line m and points A and B not on m, construct a circle on A and B which is tangent to m.
   Use a dilation this way: Let the perpendicular bisector, p, of segment AB meet m at point Z. Draw a circle with its center on p and tangent to m, and then carry out a dilation with center at Z and mapping a point of this circle to A.

6. **Construction I** Construct the circle on three given points.

7. **Construction J** Construct a circle tangent to two given lines, l_1 and l_2, and tangent to a given circle c.
   Hint: Draw lines tangent to the given circle and parallel to the given lines.

8. Draw two circles which are exterior to each other.

   (a) Construct the two centers of dilation which map one circle to the other.
   (b) Construct a line on one of the centers of dilation, a line which is tangent to one of the two circles. Now explain why that line will be tangent to the second given circle.

Institutional Transformation of Geometry: France

3

The transformations referred to in the book title are not limited to the geometric transformations of the plane, but also to the institutions in which mathematicians worked and the ways mathematics was communicated. In the late eighteenth century, this transformation was especially pronounced in France. Everything that occurred in France in the late eighteenth century was affected by the French Revolution, but it is wise to see the revolution as just the most startling event in a long period of modernization.

The medieval universities of Europe, most of which survived into the eighteenth century, taught very little science or mathematics. There were exceptions, but none in France. The universities had few students and few positions for faculty. They were poorly financed, with an ossified curriculum; some professors needed to sell alcohol or firewood to students to survive. We see changes beginning around 1620. Several elite *colleges* were founded in France, and their curriculum included mathematics and the new physics. (*Colleges* were, and are, academic high schools, some quite prestigious.)

Many of the leading mathematicians of Europe were French, including François Viète, René Descartes, Girard Desargues, and Pierre Fermat, active from the late sixteenth to mid-seventeenth century. All were independent of any academic institution. However, Descartes, Fermat, Desargues, and others were linked by a sort of institution, Father Marin Mersenne's "circle," which met regularly in Paris in the 1630s and 1640s to discuss new ideas in science and mathematics. Those who lived far from Paris, as did Fermat and Descartes, communicated by letter with Father Mersenne.

Soon afterwards, formal scientific societies began. France, under Louis XIV's minister Jean-Baptiste Colbert, made science and technology a state concern. Colbert helped found the *Académie Royal des Sciences* in 1666. There were elections of members, but appointment by the government and subsidies from the government. In England, a scientific society began meeting in 1660. In 1663 it received a royal charter as the *Royal Society of London*. At the same time we see

C. Baltus, *Geometry by Its Transformations*, Compact Textbooks in Mathematics,
https://doi.org/10.1007/978-3-031-72281-3_3

the first scientific journals. The *Philosophical Transactions of the Royal Society of London* began in 1665, and the French *Journal des Sçavants (Savants)* began the same year. In 1682 the *Acta Eruditorum* began, published in Leipzig. Its Latin name indicates its international scope.

In the seventeenth century, science became more experimental and less centered on the transmission and critique of ancient texts. Galileo (1664–1742) was just the most notable figure in that movement. The telescope and the microscope were developed during the century. In the eighteenth century, the Enlightenment swept the intellectual circles of Europe. It valued rational thinking and scientific applications. The signal project of the French Enlightenment, Diderot's *Encyclopedia*, is packed with diagrams and how-to details of industrial processes. In the 1760s, the *Académie des Sciences* issued a series of volumes which together constituted a "Dictionary of Arts and Crafts." According to Simon Schama,"These volumes were a primer not just on traditional industrial techniques but on the newest machinery" [92, p. 191].

Further, the armies of Europe came to depend on engineering skills which, in turn, depended on mathematically trained officers and printed manuals. Fortifications, artillery, and navigation all became mathematical subjects. The first school to formally train officers in these skills was founded in 1748, the *École royale du génie de Mézières*. (*Génie* means "engineering.") Admission was by rigorous oral examination, and was limited, generally, to men of noble birth. (Gaspard Monge, a commoner, gained admission by his remarkable drafting skills.) Other schools, along the same lines, were formed for the navy and artillery, and for *Ponts et Chaussées*—Roads and Bridges. All emphasized mathematics, not just for its practical applications but also as training in disciplined thinking.

The French Revolution began in 1789. Historians now tend to regard the revolution as a surface phenomenon in a broader and deeper process of modernization. "The great period of change was not the Revolution but the late eighteenth century" [92, p. 185]. The Revolution closed all the military schools. After the execution of the king, Louis XVI, in 1792, France found itself at war with several neighboring countries. Both Monge and his student at *Mézières*, Lazare Carnot, took on leadership roles in building and equipping the French army. In 1793, Monge and several others designed a new school to train engineers, both military and civilian, that were desperately needed. It was to be called the *École centrale des travaux publics*. But by the next year, the plan was revised. There was to be the *École polytechnique*, in Paris, to give instruction in science and mathematics and various mathematical applications, including fortifications, physics and rational mechanics, and, for graduates of the *École polytechnique*, three *Écoles d'application*: mines, artillery and engineering, and roads and bridges.

Due to the leading role of Gaspard Monge, the curriculum in the early years of the *École Polytechnique* included a two-year course in geometry, both analytic and synthetic.

The biggest change from the schools of the Old Regime was that noble birth was not required but, in fact, discouraged at first. Entrance requirements were initially looser than had been the case earlier, but the professors soon made entrance as rigorous as it had ever been. Another change was in the size of the institution.

There were 139 new students in 1804, for example, with four professors teaching geometry and analysis. When Monge taught at *Mézières*, he was the only master of mathematics, with one assistant. In 1804, Napoleon made the school into a military academy, with students living in barracks and always in uniform.

Unlike the scholarly academies, such as those at Berlin and St. Petersburg, to which Euler belonged, the *École Polytechnique* was primarily to train professionals, particularly military officers.

Contemporary with the establishment of the *École polytechnique* was an expansion of the institutions, in many places, that promoted mathematics. We see military academies with an emphasis on scientific education. The *United States Military Academy*, West Point, is a prime example. It was motivated by the experience of the American Revolution, but also inspired by the example of the French *École polytechnique*. It was founded by the US Congress in 1802 and expanded and reorganized by further legislation in 1812. The four year program prepared young men not only as military officers, but also, through most of its existence, as civil engineers. It was, in essence, the first engineering college in the United States.

Despite the emphasis on preparing engineers, the instructors at the *École polytechnique* were to be scholars, producing scholarly papers. Many of the papers appeared in the two publications created by the *École polytechnique*: the *Correspondance* and the *Journal*. The next generation of French mathematicians were nearly all graduates or professors of the *École polytechnique*. Its early graduates included Siméon Poisson (entered 1795), François Arago (1803), Charles-Julien Brianchon (1804), Augustin-Louis Cauchy (1805), Jean-Victor Poncelet (1807), and Michel Chasles (1812)—a who's who of a new generation in French mathematics.

In addition to publications of the *École Polytechnique*, a new journal appeared in 1810, the *Annales de Mathématiques Pure et Appliquées*, often called *Gergonne's Journal*, after its founder and first editor, Joseph Diez Gergonne. In its first decades, a large percentage of its articles were by professors or graduates of the *École Polytechnique*, most often about geometry, both synthetic and analytic. In 1826 was launched, in Germany, the publication that would soon be the most prestigious in mathematics, the *Journal für die reine und angewandte Mathematik*, or *Journal for Pure and Applied Mathematics*, called *Crelle's Journal*, also after its founder and first editor. Where the *Philosophical Transactions* of Britain's Royal Society, and the *Journal de Savants*, both started in 1665, were issued by scientific societies and covered a broad range of scientific investigations, the journals of Gergonne and Crelle were independent of particular institutions and were limited to mathematics. And the mathematicians who contributed to the journals worked in newly created departments in universities throughout Europe, universities open to far more students than ever before.

Affinity and the List of Transformations by Moebius

4

1   Transformations as Listed by Moebius

Geometric "transformation" was first given specific attention as a thing in itself in *Der Barycentrische Calcul* [67], of 1827, by August Ferdinand Moebius (1790–1868). Where we refer to a "transformation," Moebius used the German name *Verwandtschaft*, a word for "relationship." In practice, the nineteenth century use was often closer to a mathematical relation than a function.

First in Moebius's list of transformations are *Gleichheit* (equality or congruence) and *Aehnlichkeit* (similitude). These are familiar transformations, with the dilation a special case of similitude. We shall just give definitions, following Moebius, then move to the third of his listed transformations, the *Affinität*, *affinity*. The fourth for Moebius was *collineation*. In the projective plane, collineation is the general *projective transformation*, the topic of several coming chapters. In the chapter on homogeneous coordinates, we will present Moebius's proof that a collineation of the projective plane is determined by four points, no three collinear, and their images.

Here are the definitions.

Definitions

Two figure are *similar*, and the relation is called a *similitude*, pairing each point X of the first figure with one point X' of the other, if the relation respects distance. This means there is a positive constant k so the distance between any pair of points X and Y of the first figure is k times the distance between the corresponding pair X' and Y' of the second figure.

When the distances are equal, the relation is an *equality* or *congruence*.

(continued)

© The Author(s), under exclusive license to Springer Nature Switzerland AG 2025
C. Baltus, *Geometry by Its Transformations*, Compact Textbooks in Mathematics,
https://doi.org/10.1007/978-3-031-72281-3_4

> A *collineation* is a transformation that pairs collinear points with collinear points.
>
> An *affinity* is a collineation in which parallel lines are paired with parallel lines.

A *dilation* is an example of an *affinity*. We most often see affinities in the real plane, $\mathbb{R}^2$. In the projective plane, for example, where any two lines meet in a point, an affinity makes no sense. $\mathbb{R}^2$ is an *affine plane*, the only case we will deal with. It has these characteristic properties:

In an *affine plane*

(*i*) on any two points there is exactly one line, and
(*ii*) for any line m and point X not on m there is exactly one line parallel to m.

2   Affinity According to Moebius

Affine transformations, applied to curves, appeared in an 1748 work of Euler [40, Part II, Chapter 18]. If there are constants p and q, and all points (x, y) of one curve are replaced by (px, qy) to create a second curve, then the curves were called *affine*. Euler's description fits into Moebius's concept, but Moebius had a broader concept. His first characterization involves three particular finite points, A, B, C, which are not collinear, and their respective finite images, also not collinear, A', B', C' [67, p. 193]. In this case, any point P can be represented as $P = pA + qB + rC$.

> **Definition**
> Suppose non-collinear finite points A, B, and C are to be mapped, respectively, to non-collinear finite points A', B', and C'. Then an *affinity* is defined by mapping a point $P = pA + qB + rC$ to $P' = pA' + qB' + rC'$. All affinities of $\mathbb{R}^2$ can be defined this way.

Moebius offered an alternate characterization: an affinity is a collineation so that when a segment AB is cut by a point X in a certain ratio, then the image segment is cut by X' in the same ratio [67, p. 199]. We will see in the matrices chapter why this is reasonable in terms of a modern understanding of an affine transformation.

3   Application: The Ellipse as the Image of a Circle Under an Affinity

The ellipse $\dfrac{x^2}{a^2} + \dfrac{y^2}{b^2} = 1$ is the image of circle $x^2 + y^2 = a^2$ under the affinity that replaces y by ay/b. (We called such a transformation a *vertical stretch* by *scale factor b/a* in that any point (x, y) on a given curve is replaced by $(x, by/a)$.)

We apply this transformation to a circle to prove Prop. 31 of Book 7 of the *Conics* of Apollonius. Isaac Newton used this proposition as Lemma 12 of Section 2 in his *Principia Mathematica* [73], of 1687, and we have kept the notation of Newton's work.

The theorem involves *conjugate diameters*. As we noted with Apollonius's *Conics*, a *diameter* of a conic is the set of midpoints of all the chords of the conic which are parallel to a particular chord. The chords corresponding to a particular diameter are its *ordinates*. If one diameter of a conic is an ordinate for another diameter of the conic, then those diameters are *conjugate*. In a circle, perpendicular diameters are *conjugate*.

Theorem 4.1 *In an ellipse, take diameter GP and its conjugate diameter DK. Then the parallelogram formed by the tangents at D, P, K, G has the same area no matter the choice of the conjugate diameters.*

A *vertical stretch* has these properties. ((i) and (ii) are properties of all affine mappings):

(i) parallel segments are mapped to parallel segments,
(ii) the midpoint of a segment is mapped to the midpoint of the image segment, and
(iii) the area of a figure is multiplied by the scale factor of the vertical stretch.

Proof To see this last property, we assume the figure can be approximated in its area as closely as desired by inscribed horizontal rectangles of height δ. The height of each such thin rectangle is multiplied by the scale factor while its length is unchanged by the vertical stretch. Thus the area of each rectangle is multiplied by the scale factor, so the area of the entire figure is, likewise, multiplied by the scale factor.   □

To prove the theorem for ellipse $\dfrac{x^2}{a^2} + \dfrac{y^2}{b^2} = 1$, we consider the circle $x^2 + y^2 = a^2$. The conjugate diameters GP and DK, as in Fig. 4.1, are the images of conjugate diameters $G'P'$ and $D'K'$, respectively, of the circle. In a circle, conjugate diameters are perpendicular. With the circle, the parallelogram formed by the tangents at D', P', K', G' is a square of area $4a^2$, a value that is the same for every pair of perpendicular diameters. Therefore, in the ellipse, the corresponding

Fig. 4.1 Proof of Newton's
Section 2 Lemma 12, 1687

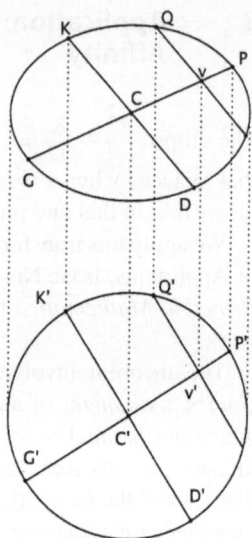

parallelograms have the same area no matter which pair of conjugate diameters are considered.   □

4   Exercises—Affinity and Transformations as Listed by Moebius

1. Prove that the composition of two affinities is itself an affinity.
2. By applying an affinity to the circle $x^2 + y^2 = a^2$, find the area of the ellipse $\dfrac{x^2}{a^2} + \dfrac{y^2}{b^2} = 1$.
3. Prove that with an affine plane, a collineation transformation is an affinity.

Background for Homology: The Common Secant, the Cross-Ratio, and Harmonic Sets

5

The next transformation we will examine is *homology*, a special case of the more general *projective transformation*. Poncelet's development of homology, a process we can trace from his *Notebooks* of 1813–1814 through his masterwork, the *Traité* of 1822, involves several important topics. We present these one at a time, beginning with the *common secant*.

1   Common Secant (or Chord) of Two Circles

We have already noted the Common Secant Theorem, as stated and proved by Lazare Carnot: when three circles meet pairwise, then the three (pairwise) common secants are concurrent. In Poncelet's first *Cahier* of 1813, written in a military prison in Russia, he extended the idea of a common secant to a pair of circles that do not meet. As it turns out, back in France, another graduate of the *École Polytechnique*, Louis Gaultier, developed the same idea in a very different way, and he called the common chord the *radical axis* of a pair of circles [46]. This expanded Common Secant Theorem, which we will see below, is Proposition 8 of Poncelet's *Cahier* 1.

Students may have encountered already this idea in working with equations of circles. Let us find the common chord of circles that meet, say circles $x^2 + y^2 + 4x = 9$ and $x^2 + y^2 - 8y = 4$. Subtracting one equation from the other produces the equation $4x + 8y = 5$, the equation of a line. What line is it? The line on the two points at which the circles meet. In other words, $4x + 8y = 5$ is the equation of the common chord or common secant of the two circles. Now, if the two given circles do not meet, the same steps produce the equation of a line. This line will lie between the two circles when neither circle lies inside the other. Poncelet called it the *common chord* or *secant* of the two circles. It will turn out to be an important step in Poncelet's development of projective geometry.

Poncelet found a helpful way of characterizing the common secant of two circles, a characterization which holds even when the circles do not meet.

Fig. 5.1 Tangents from a point on the common secant to the two circles are equal

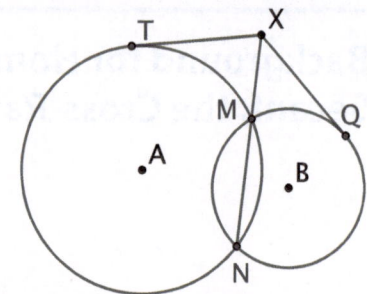

Fig. 5.2 MD is the *Common Chord* of circles A and B

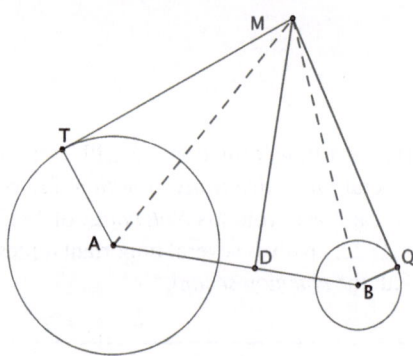

To start, the reader is asked to show that when two circles, with centers A and B, meet in points M and N, then tangents to the two circles from any point on MN and outside the circles have equal length. See Fig. 5.1. Suggestion: Consider the power of point X with respect to the circles.

> **Definition**
> 1. (Poncelet) Given two circles, not concentric, their *common chord* or *secant* is the line on the set of points from which tangents to the circles are of equal length.
> 2. (Steiner) Given two circles, M and m, not concentric, their *common chord* or *secant* is the set of points of equal power with respect to the two circles. Steiner called that set the "line of equal powers of circles M and m [96, p. 165]."

For Poncelet's definition to have meaning, we need to show that this set of points forms a line. The claim is this theorem. See Fig. 5.2.

Theorem 5.1 *Suppose there are equal length tangents, MT and MQ, from a point M to circles with centers A and B and radii r_A and r_B. Then the perpendicular from M to line AB meets AB at a point whose location depends only on A, B, r_A and r_B, and not on the choice of M.*

Proof *(Modeled on Poncelet's [84, Cahier 1 Prop. 6] of 1813)* Let the perpendicular from M meet AB at D. Draw segments MA, MD, and MB. By the Pythagorean Theorem,

$$TM^2 + r_A^2 = AM^2 = MD^2 + AD^2.$$

$$\text{Likewise,} \quad QM^2 + r_B^2 = BM^2 = MD^2 + BD^2.$$

Subtract the equations and note that $AD + DB = AB$ and $TM = QM$, to see that AD depends only on A, B, r_A and r_B. So all such points M lie on the perpendicular to AB on the same point D.   □

With this theorem, we see that points of equal power to two given circles and lying outside the circles lie on the common secant of the circles. Earlier, in the Points of Equal Power Theorem, we saw that points inside two given circles and of equal power with respect to the two circles are on the common secant of the two circles. Thus we have an alternative definition of the *common secant* of two circles, due to Jacob Steiner, from 1826. The Common Secant Theorem follows immediately since the point at which the common secant of circles c_1 and c_2 meets the common secant of circles c_2 and c_3 has equal power with respect to circles c_1 and c_3. Recall that for a point outside a given circle, the power of the point is the square of the length of the tangent to the circle.

Theorem 5.2 (Common Secant Theorem) *Given three circles, the three pairwise common secants of the circle are concurrent. The point of concurrency is sometimes called the* radical center *of the circles.*

A helpful construction to locate the common secant of two circles, from Jacob Steiner's 1826 [96, Article 4], is based on the Common Secant Theorem. The construction is described in the following theorem; the proof is an Exercise.

Construction K Steiner Common Secant Construction.

Theorem 5.3 *For circles C_1, C_2, draw a circle c, meeting circle C_1 in A_1 and B_1, and meeting circle C_2 in A_2 and B_2. Then secants $A_1 B_1$ and $A_2 B_2$ meet on the common secant of C_1, C_2.*
(The common secant is perpendicular to the line on the centers of the two circles.)

---

2   Harmonic Conjugates, the Cross-Ratio, and Their Invariance

It was recognized that plane-to-plane and line-to-line projection preserved certain features of geometric figures. Two such features are closely related: a "line divided harmonically," as Philippe de la Hire expressed it in 1673 [60], and, coming later, the *cross-ratio* of four collinear points.

Here are their definitions.

Definitions

If four points A, B, C, D on a line satisfy

$$AD \cdot CB = AB \cdot CD, \text{ with exactly one of } B \text{ and } D \text{ between } A \text{ and } C,$$
$$(5.1)$$

then A and C are called *harmonic conjugates* of B and D, and the statement is abbreviated $H(AB, CD)$. When the order of the four points is understood, we will simply say the points are "harmonically related," and that the set is a "harmonic set." If the lengths in the equation are understood to be signed lengths, then the condition on the order of the points is necessarily satisfied.

Let A, B, M, and N be collinear points. Then the *cross-ratio*, denoted $CR(AB, MN)$, is

$$CR(AB, MN) = \frac{AM \cdot BN}{AN \cdot BM},$$

where the pairs of letters indicate a signed distance.

The cross-ratio of harmonically related points, in the appropriate order, is equal to -1. We further note that the cross-ratio and harmonic conjugates may involve infinity. We find the value as the limit as one of the points approaches infinity, a concept especially important in a harmonic set. The main points are in the following theorem.

Theorem 5.4

(i) *Points A and C are harmonic conjugates of B and D, i.e., $H(AC, BD)$, exactly when $CR(AC, BD) = -1$.*

(ii) *The midpoint of any segment is the* harmonic conjugate of ∞.

It is a major theorem that the projected images of harmonically related points are themselves harmonically related. Desargues called this arrangement of points a *four point involution*, a special case of *involution*. He gave a proof that involution is preserved under projection. More generally, the cross-ratio of collinear points is preserved under a projection. In Fig. 5.3, we have collinear points A, B, M, N projected, respectively, to collinear points A, P, C, Q.

We give a proof of invariance of the cross-ratio under projection, implying that a harmonic set of points is projected to a harmonic set of points. This proof is based on the idea employed by Carnot in his 1806 proof that the harmonic relation is invariant [22, p. 93].

Fig. 5.3 Proof of invariance
of the cross-ratio, based on
Carnot's [22], 1806

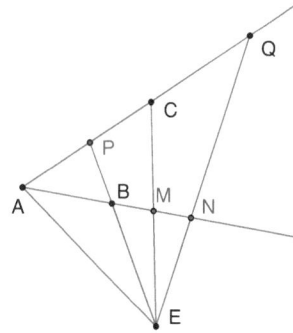

Theorem 5.5 (Invariance of the Cross-Ratio) *Let two line meet at A, with points
A, B, M, N on one line projected from point E to, respectively, A, P, C, Q. Then*
$CR(AB, MN) = CR(AP, CQ)$.

As a corollary, since a segment AD is divided harmonically *at B and C exactly
when $CR(AC, BD) = -1$, then the harmonic relation is invariant under a line-to-
line projection. (When the two lines involved do not meet at one of the four points,
we can draw an additional line so that invariance of the cross-ratio follows from
two applications of the claim of the theorem.)*

Proof See Fig. 5.3. (First, assume segment lengths are positive.) By the Law of
Sines for $\triangle EBM$ and $\triangle EBN$,

$$\frac{sin\angle BEM}{sin\angle EBM} = \frac{BM}{ME} \quad \text{and} \quad \frac{sin\angle BEN}{sin\angle EBN} = \frac{BN}{NE}.$$

Since $\angle EBM \cong \angle EBN$, it follows that

$$\frac{BN}{BM} = \frac{sin\angle BEN}{sin\angle BEM} \cdot \frac{NE}{ME}.$$

In the same way, we see that

$$\frac{AM}{AN} = \frac{sin\angle AEM}{sin\angle AEN} \cdot \frac{ME}{NE}.$$

In multiplying these last two equations, ME and NE vanish, so that cross-ratio
depends only on the angles that lines AE, BE, ME, NE form at E:

$$\frac{AM \cdot BN}{AN \cdot BM} = \frac{sin\angle AEM}{sin\angle AEN} \cdot \frac{sin\angle BEN}{sin\angle BEM}.$$

So we must have equal cross-ratios: $\dfrac{AM \cdot BN}{AN \cdot BM} = \dfrac{AC \cdot PQ}{AQ \cdot PC}.$ □

The sign of the cross-ratio depends on the order of the four points. Check in the two cases, where E is between lines AQ and AN and when it is not, that the sign of the cross-ratio is unchanged.

Charles-Julien Brianchon (1783–1864), in [19] of 1817, claimed without proof that the cross-ratio is invariant under projection. A lemma guaranteeing invariance goes back to Pappus [76, Book 7 Prop. 129]; invariance also follows from Poncelet's Art. 20 of [87] of 1822. Moebius examined the cross-ratio in [67] of 1827 where, in Art. 188, he proved its invariance under projection.

Note that three collinear points in a given order and a given value of the cross-ratio (not 0, 1 or ∞) uniquely define the fourth collinear point. In the same way, the fourth point of a harmonic set, for a given ordering of points, is determined. Here is the statement, with proof in the case of a harmonic set.

Theorem 5.6

(i) *When point D, collinear with A and C, is not between A and C, then there is a unique point B between A and C where $H(AC, BD)$.*

(ii) *Given collinear points A, B, and C, and a value k (not 0, 1 or ∞) , then there is a unique collinear point D so $CR(AC, BD) = k$.*

Proof For invariance of the harmonic relation, when points A, C, and D are ordered $A - C - D$ and $AB = x$, then equation

$$\frac{AB}{BC} = \frac{AD}{CD} \text{ is } \frac{x}{AC - x} = K,$$

where K is a constant, has a unique solution. When the order of the points is $A - C - D$, we can check that $0 < x < AC$. □

---

3   Dual Statements and the Cross-Ratio

We form the *dual* of a statement about lines and points by interchanging the words "line" and "point," and interchanging "point X lies on line m" and "line x lies on point M." In projective geometry, the duals of true statements are themselves true. So we expect to have a cross-ratio of four concurrent lines and a harmonic set of concurrent lines.

The invariance under projection of the cross-ratio and of harmonic sets has its dual, leading to the defining property of harmonic conjugates among concurrent lines and of the cross-ratio of four concurrent lines.

Definition

Let a, b, c, d be four lines concurrent at a point E.

We define the *cross-ratio* $CR(ab, cd)$ to be $CR(AB, CD)$ where A, B, C, D are the four points at which some line not on E meets lines a, b, c, d, respectively.

Likewise, lines a and b are *harmonic conjugates* of lines c and d when A and B are harmonic conjugates of C and D, when some line not on E meets lines a, b, c, d, respectively, in A, B, C, D.

Note from the proof above of the invariance of the cross-ratio that the cross-ratio of concurrent lines *depends only on the angles formed by the lines*.

Cross-ratio invariance leads us to a helpful theorem. Philippe de la Hire proved it as Lemma 7.1 in the case of a harmonic set of points in [60, 1673]; Jacob Steiner proved it in the more general case of the cross-ratio in 1832 [98, Art. 9 p. 259].

Theorem 5.7 (La Hire-Steiner Theorem) *When two lines meet at D, and $CR(AB, CD) = CR(A_1B_1, C_1D)$ for points A, B, C of one line and points A_1, B_1, C_1 of the second line, then lines AA_1, BB_1, and CC_1 are concurrent. We get the corresponding conclusion for harmonic sets.*

Proof Let S be $AA_1 \cap BB_1$; then line SC must meet the second line in the one point C_1 that makes $CR(A_1B_1, C_1D)$ equal to $CR(AB, CD)$. □

4 Carnot, the Complete Quadrilateral, and Harmonic Conjugates

Lazare Carnot highlighted the concept of the *complete quadrilateral* and gave it that name [21, p. 275]. It is an important theorem that a complete quadrilateral generates a harmonic set. The proof below is from Carnot in 1806 [22].

Definition A *Complete Quadrilateral* is a set of four lines, a, b, c, d, no three concurrent, ordered as consecutive sides.

Theorem 5.8 (Carnot, Theorem 1.6, 1806) *Let A, B, C, D be the four vertices of a quadrilateral, giving sides AB, BC, CD, AD of a complete quadrilateral (no three concurrent).*

Fig. 5.4 Complete
Quadrilateral $ABCD$ yields
harmonic conjugates:
$H(LM, JK)$

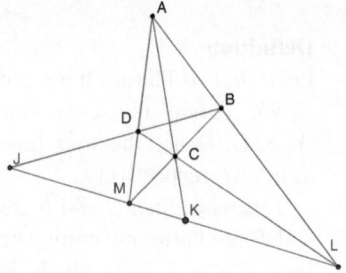

Let $AB \cap CD = L$ and $AD \cap BC = M$. Let diagonal AC meet line LM at K and let the diagonal BD meet line LM at J. Then L and M are harmonic conjugates of J and K.

Proof See Fig. 5.4. We apply Ceva's Theorem to $\triangle LMA$, where the lines (*cevians*) on the three vertices meet at C. This gives $DM \cdot BA \cdot KL = DA \cdot BL \cdot MK$. Then apply Menelaus's Theorem to $\triangle LMA$ cut by line JDB. This gives $JM \cdot DA \cdot BL = JL \cdot DM \cdot BA$. Equate the product of the left sides of the two equations to the product of the right sides, and simplify. The result is $JM \cdot KL = JL \cdot KM$. We also note that exactly one of J and K is between M and L. So $H(LM, JK)$.   □

Here is an elegant modern proof. See Fig. 5.4. Let $DB \cap AC$ be E. By projection from A, $CR(JK, ML) = CR(JE, DB)$. By projection from C, $CR(JE, DB) = CR(JK, LM)$. Now, $CR(JK, ML)$ and $CR(JK, LM)$ are reciprocals. The only numbers equal to their reciprocals are 1 and -1. By the order of the points, $CR(JK, ML) = -1$. Therefore, $H(JK, ML)$.   □

Given any harmonic set, we can create a complete quadrilateral that produces the given harmonic set, as in the theorem above. A collineation maps a complete quadrilateral to a complete quadrilateral. This theorem follows:

Theorem 5.9 *Any collineation maps a harmonic set to a harmonic set.*

5   Conic Sections, Pole and Polar, and Harmonic Sets

We have noted that a conic section, as defined by Apollonius, is a slice of a conic surface by a plane. See Fig. 5.5. Someone attuned to plane-to-plane projection, as was Girard Desargues, in the 1630s, and Philippe de la Hire, a generation later, will see the conic section as the projection, from the vertex of the cone, of the base circle onto the slicing plane. (It may be significant that Desargues had written a pamphlet on perspective drawing and La Hire was himself an artist.) We note that the diameter, ED, of the conic is the projection of a diameter of the base circle.

Fig. 5.5 From the 1696
edition of the *Conics* of
Apollonius [5]

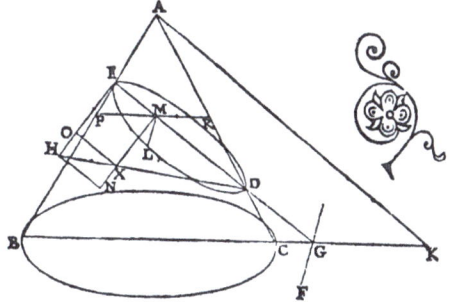

> **Definition** A *conic section* is the image of a circle under a plane-to-plane
> projection.

Pole and Polar Here we introduce the terms *pole* and *polar* with respect to a conic
section. The basic concept of *pole* and *polar* was well understood by Desargues and
la Hire in the seventeenth century. *Pole* and *polar* concepts, in various special cases,
are found in the *Conics* of Apollonius. That the special cases were not gathered
under one general concept, and that no name was given for *pole* and *polar*, suggest
limited understanding and/or interest on the part of Apollonius. Desargues had a
name for the *polar* of a point, a *traversal*. The first clear enunciation of the concept,
together with the first use of the related names *pôle* and *polaire*, was by J. D.
Gergonne in the 1812–1813 volume of his *Annales* [47]. Gergonne developed the
pole-polar relationship by the second degree equations for a conic.

> **Definition** Let A be a point outside a conic section, with tangents from A
> meeting the conic at K and L. Then line KL is the *polar* of A (with respect
> to the conic section) and A is the *pole* of line KL. (See Fig. 5.6.)

Apollonius showed in Book 1 Props. 17, 32, and 37, and in Book 3 Prop. 37,
how the "harmonic division of a line," as Philippe de la Hire put it in [60, 1673],
appeared with the conic sections. We summarize in this theorem:

Theorem 5.10 (*Conics* **of Apollonius**) *(See Fig. 5.6.) Let A be a point outside a
given conic, and let ordinate KL be the polar of A. Let diameter ED, meeting KL
at its midpoint M, be the line made of the midpoints of all the chords parallel to
KL. Then*

   (i) A lies on ED and $H(AM, ED)$, and
   (ii) the tangent at E is parallel to ordinate KL.

Fig. 5.6 *A* is the Pole of line
KL.
$H(AM, ED)$, $H(AG, HN)$,
and $H(BG, KL)$. The
chords, such as *KL*, parallel
to each other and bisected by
ED are *ordinates* with
respect to *diameter ED*

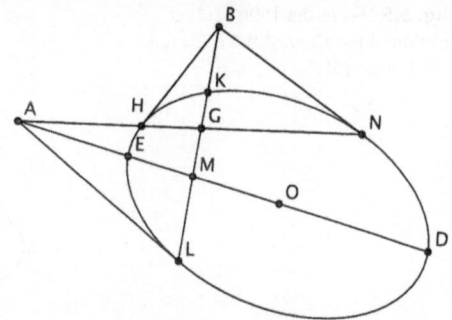

(iii) *Let another secant on A meet the conic at H and N and cut the polar at G.*
Then H(AG, HN), and tangents at H and N meet on the polar LK.

In this way the pole-polar relationship is related to the harmonic relationship.
Apollonius had proved (*i*) and (*iii*) in separate cases for the three types of conics.
This claim was proved in the seventeenth century by Desargues, in [35, 1639].
(*i*) and (*iii*) will be proved in a theorem below. Our proof, largely covered in our
Lemmas 5.1 and 5.2, follows the 1673 work of Philippe de la Hire.

Desargues and La Hire proved the theorem in the case of a circle. Then, as La
Hire pointed out, because the harmonic relationship is invariant under projection
and a conic section is a projection of a circle where the diameter of the conic is the
projection of a diameter of base circle, the claim holds for any conic section.

For (*ii*), we offer a simpler proof than those of Apollonius and La Hire.

Proof Fixing the conic and diameter *ED*, let *A* approach *E* along the diameter,
so point *L* approach *E*. Then the ordinate on *L* approaches the tangent at *E*. By
the limiting case, tangents at *E* and *D* are parallel to the ordinates with respect to
diameter *ED*.   □

Lemma 5.1 (La Hire's Lemma 8 of 1673) See Fig. 5.7. *In La Hire's Fig. 12, FG*
is the polar of point A, and the diameter on A is drawn, cutting the circle at B and
E and cutting the polar at C. Tangents are drawn at B and E. Then AE is divided
harmonically at B and C.

Proof By similar triangles, noting $LF = LB$ and $HF = HE$, we arrive at $AH \cdot LF = HF \cdot LA$. By a parallel projection onto the diameter, we get $AE \cdot BC = CE \cdot AB$.   □

We next consider a secant on *A* that is not a diameter,

Lemma 5.2 (La Hire's Lemma 9 of 1673) See La Hire's *Fig.* 13 in our Fig. 5.7
Left. *Let A lie on diameter BE extended of circle BEG, and let a secant, not a*

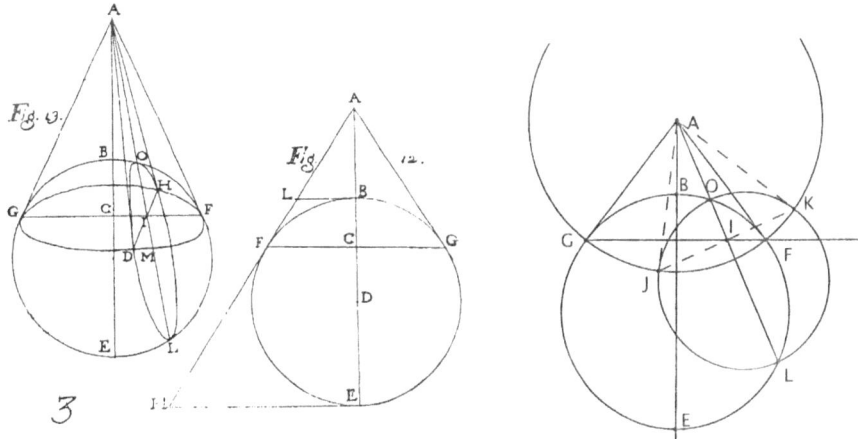

Fig. 5.7 Left and Center: La Hire's 1673 *Fig.* 13 and 12. Secant dividing a circle harmonically. Right: Alternate proof of La Hire's Lemma 9 of 1673

diameter, on A meet the circle at O and L and meet the polar, GF, of A at I. Then AL is divided harmonically at O and I.

Proof *(La Hire)* Take the sphere for which circle EBG is a great circle and slice it along line AL perpendicular to the plane EBG. That plane slices the sphere in a circle OHL for which AL is a diameter. So by La Hire's Lemma 8 of 1673, $AL \cdot OI = AO \cdot IL$. □

Here is an alternate proof, in two dimensions [8], similar to what la Hire offered in 1685.

Proof In Fig. 5.7 Right, we draw circle $OKLJ$ with diameter OL, where JK is the polar of A with respect to the circle OJL. We draw the circle on J and K with center A. Because AL is a common secant to circles BGE and OJL, by the power of point A, $AG = AF = AJ = AK$. So G, J, F and K lie on a circle with center A. By the Common Secant Theorem, JK, OL, and GF are concurrent. So I, on GF, is $JK \cap AL$. This means AL is divided harmonically at O and I. □

Let us take up an interesting construction, by straightedge only, of a tangent to a conic section from an outside point.

Construction L Construct the polar of an outside point, A.

See Fig. 5.8. On point A, draw two secants to the conic, one cutting it at E and F, the second at G and H. The polar we seek cuts the first secant at I and the second secant at J. Since $H(AI, EF)$ and $H(AJ, GH)$, and $H(AI, FE)$ and $H(AJ, GH)$, then

Fig. 5.8 Straightedge
construction of tangents from
A to a conic section

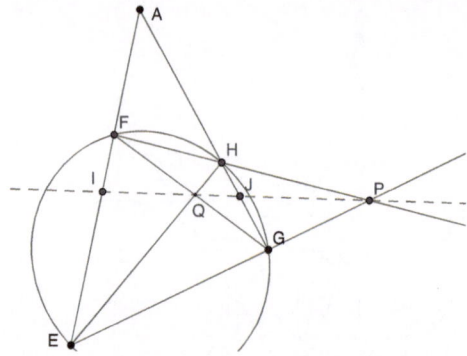

by the La Hire-Steiner Theorem, the polar of A is concurrent at P with lines FH
and EG, and at Q with FG and EH. So the tangents from A meet the conic section
where line PQ meets the conic.   □

As an additional claim about Fig. 5.8, we show that the tangents at H and G meet
on the polar of A.

Theorem 5.11 *Let a secant on point A outside a conic section meet the conic at H
and G. Then the tangents at H and G meet on the polar of A.*

Proof In Fig. 5.8, fix A and secant AHG, and let the secant AFE turn about A. As
F approaches H and E approaches G, lines FH and EG approach the tangents to
the circle at H and G, respectively, all the while still meeting on the polar of A. In
the limit, we have the theorem.   □

To complete the pole-polar concept, we need to define the polar of a point inside
a conic section. Recall that Monge's Pole-Polar Theorem shows that a line m outside
a conic has its pole, M, inside the conic, where each point of m has its polar on M.

> **Definition**
> For a line m meeting a conic at points E and F, the *pole* of m is the point at
> which tangents at E and F meet (outside the conic).
>
> For a point M inside a conic, it follows from Monge's Pole-Polar Theorem
> that all lines on M have poles on a line m outside the conic. (See Exercise 5.)
> That line m is the *polar* of M.
>
> For a point M on a conic, its *polar* is the tangent to the conic at M.

The following theorem about poles and polars can be summarized by saying, "A
point X lies on line z exactly when point Z lies on line x."

Theorem 5.12 (Pole-Polar Theorem) *Given a conic section, there is a one-to-one pairing of each point X, except the center of the conic, with a line x, called the polar of X, with this property: a line z is on X exactly when Z, the pole of z, is on x. Under such a pairing, X is called the* pole *of x.*

When X is outside the conic and E and F are the two points of contact of the two tangents from X to the conic, then line EF is the polar *of X (with respect to the conic) and X is the* pole *of line EF.*

When X is inside the conic, the poles *of the lines on X form a line, the* polar *of X.*

When X is on the conic, the tangent at X is its polar.

(In projective geometry, the center of a conic and the line at infinity are pole and polar, and this relationship holds when the conic is projected to another conic section.)

6   Exercises—Background for Homology

1. Find the common secant (radical axis) of circles $y^2 = x(4 - x)$ and $x^2 + y^2 = 2x - 6y - 9$.
2. We are given two circles of different sizes, c_1 and c_2, which meet in two points. Let A_1 be a point of c_1 which is outside circle c_2. Find point A_2 of circle c_2 so there is a third circle tangent to c_1 at point A_1 and tangent to circle c_2 at A_2. (This problem will appear again in the Exercises at the end of the Geometric Inversion chapter.)
   Suggestion 1: Let the common external tangents meet at point S. Use propositions of Poncelet's *Cahier* 1 to pair A_1 with A_2.
   Suggestion 2: Common Secant Theorem.
3. Prove the Steiner Common Secant Construction: For circles C_1, C_2, draw a circle c, meeting circle C_1 in A_1 and B_1, and meeting circle C_2 in A_2 and B_2. Then secants $A_1 B_1$ and $A_2 B_2$ meet on the common secant of C_1, C_2.
   Hint: Apply the Common Secant Theorem.
4. Let $A = (-4, 0), B = (-2, 0), C = (1, 0), D = (4, 0)$. Project the four points onto line $y = -4$ through center $(-1, -1)$. Show that the cross-ratio $CR(AB, CD)$ is unchanged by the projection.
5. Let M be a point inside a conic and AB, CD, and EF chords of the conic which lie on M. Let tangents at A and B meet at X, tangents on C and D meet at Y, and tangents on E and F meet at Z. Prove that X, Y, and Z are collinear.
   Hint: Monge's Pole-Polar Theorem.

Plane-to-Plane Projection

6

1   Preparing for Plane-to-Plane Projection

We now consider the *homology* transformation, introduced and named by Poncelet.

Figure 6.1 displays the homology mapping points of plane π to the points of plane π', making use of a point S which lies on neither plane. S is called the *center*. Point A of plane π is paired with the point A' of plane π' that lies on line SA. The points of the line of intersection of the two planes are all mapped to themselves; in other words, it is a *line of fixed points*, called the *axis*. A special line is that pictured on point X of plane π, the line of intersection of π and the plane on S that is parallel to plane π'. The points of that line have no (finite) images in plane π'. So that line on X is called the *vanishing line* of plane π, and we think of its image as the *line at infinity* of plane π'.

Girard Desargues was the first, in his *Brouillon project d'une atteinte aux événements des rencontres d'un cône avec un plan*, of 1639, to find a rich collection of properties that are preserved under projection and to apply his study to conic sections. That 1639 work will be referred to as *Brouillon project*. The full title can be translated as *Draft project investigating the intersection of a plane with a cone*.

Although the *Brouillon project* was lost for 200 years, a couple of its developments had an impact before the book itself was recovered in 1845. First, parallel lines in the plane were regarded as concurrent at a point at infinity, an idea that stayed in the air until it was fully accepted as part of projective geometry early in the nineteenth century. Second, Desargues introduced the *involution* relation, which we will see later in this book.

Poncelet discovered homology, the main subject of his 1822 *Traité*. What from the previous centuries that led to his discovery?

First, there are notions of plane-to-plane mappings in three-dimensional space, beginning with painting in fifteenth-century Italy. Rules to represent a tile floor on a canvas began with Leon Battista Alberti, in 1435 [4]. These are rules of *perspective drawing*. A point X on a tile floor was to determine a point X' on the plane of a

© The Author(s), under exclusive license to Springer Nature Switzerland AG 2025
C. Baltus, *Geometry by Its Transformations*, Compact Textbooks in Mathematics,
https://doi.org/10.1007/978-3-031-72281-3_6

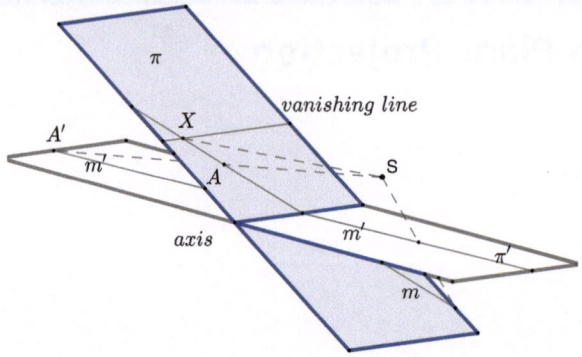

Fig. 6.1 Projection from point S of plane π plane π'

Fig. 6.2 Based on 'sGravesande, 1711, $Fig.$ 5, projection of CD onto cd. M is above c and d

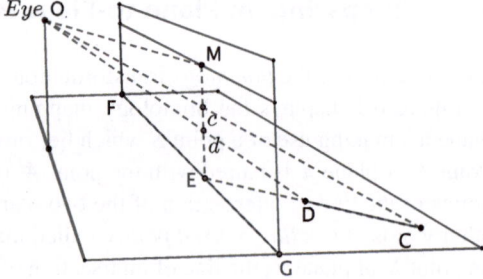

canvas, where X' was the point at which the line on X and the painter's eye, O, met the plane of the canvas.

See Fig. 6.2, based on a Willem 'sGravesande's $Fig.$ 5 of 1711, showing the eye at point O [50]. In the diagram, the rays joining segment CD to the eye at O meet the vertical plane of the canvas in segment cd.

Note that the lines CD and cd meet in a point E. We summarize properties of this plane-to-plane mapping in a theorem.

Theorem 6.1 *A plane-to-plane projection, from a point O on neither plane, is a collineation in that a line is mapped to a line. Further, a line of the first plane and its image meet on the line of intersection of the two planes.*

Proof See Fig. 6.2. Any line in the floor plane, such as line CD, determines, with O, a plane in space. Any two planes not parallel meet in a line, so plane CDO meets the vertical plane MEG in a line, cd. Of the three planes: plane CDO, the plane of the floor, and the vertical plane, no two are parallel, so they meet in a single point.   □

A further observation is that each family of parallel lines on the floor corresponds, on the plane of the canvas, to a family of lines converging at a single point, such as M, on the *horizon line* or *vanishing line*.

La Hire went a step further in 1673. Instead of projecting a circle in the base plane to a conic section in a slicing plane, the circle and conic may lie in a single plane. We will further describe the work of la Hire in connection with homology.

Another early developer of transformations was Gaspard Monge. Monge's *Géométrie Descriptive*, which he began teaching around 1770, speaks of "la méthode de projections." He described a line in space by its orthogonal projections onto two perpendicular planes. (*Orthogonal projection* onto a plane means perpendicular to that plane.) If, for example, a line in space is projected orthogonally onto the plane $x = 0$ as line $2y - z = 4$, that means our original line lies in plane $2y - z = 4$. And if the same line is projected orthogonally onto plane $y = 0$ as line $x + 3z = 1$, the line lies in plane $x + 3z = 1$, and so the original line is simply the intersection of those two planes.

2 Projecting a Line to Infinity: Brianchon, Poncelet, and Desargues' Theorem

Poncelet, in his *Traité* of 1822, wrote that he owed to Charles-Julien Brianchon, in a work of 1810, "the first idea of my work." [87, p. $xxxiv$] That idea was to project the plane of a figure onto another plane so that a particular line, m, was mapped to the line at infinity. That way, lines in the original figure which meet on m are mapped to parallel lines. (One problem Brianchon solved is presented in Application 1, at the end of the chapter.)

Brianchon, too, had his predecessors. An 1807 problem by Loius Poinsot is presented in Exercise 1, with his solution in the Solutions section.

Now let us look at Desargues' Theorem [36] and a proof employing a plane-to-plane projection in which a particular line is mapped to the line at infinity. We follow Poncelet's proof, from his 1822 *Traité*, Art. 167. (Poncelet may have been the first in the nineteenth century to attribute the theorem to Desargues.)

Theorem 6.2 (Desargues' Theorem, 1648) *Suppose two triangles are in perspective from a point, i.e., there is a correspondence of vertices of the triangles so the three lines on corresponding vertices are concurrent, then the three points of intersection of corresponding sides are collinear. And the converse holds.*

Proof (Poncelet) See Fig. 6.3. We are given triangles BLM and DNP where the lines joining corresponding vertices, PM and NL "will meet in H on the diagonal $[BD$ of quadrilateral $ABCD]$."

We must show that I, where sides NP and ML meet, lies on line AC, where $A = BL \cap DN$ and $C = BM \cap DP$. We proceed as Poncelet did in Art. 167,

Fig. 6.3 Desargues'
Theorem, from Poncelet,
1822, based on Art. 166 and
167

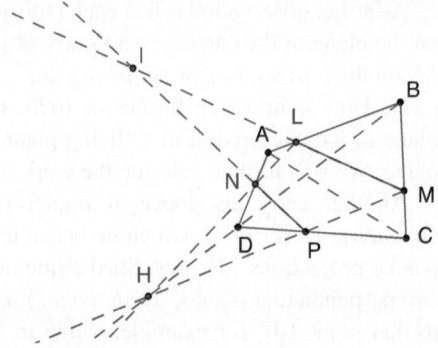

placing the figure by projection on a new plane, so the diagonal AC passes to infinity . . .
the triangles BLM, DNP necessarily change into similar triangles similarly placed [with
corresponding sides parallel] and having point H as the *center of similitude* or point of
concurrence which joins, in pairs, the homologous vertices.

"Homologous" means corresponding. Since A and C now lie at infinity, $BL \parallel$
ND and $BM \parallel PD$. Since the triangles are in perspective from H, their third sides
are also parallel: $ML \parallel NP$. Thus, $ML \cap NP$ is collinear with A and C, on the line
at infinity. So in the original figure, the corresponding sides of the two triangles in
perspective meet in three collinear points.   □

Problem (See Exercise 2 at the End of the Chapter) Prove the claim that, after
line AC is projected to the line at infinity, triangles BLM and DNP are similar.

---

3   Hexagon Theorems of Pascal and of Brianchon

Blaise Pascal (1623–1662) produced major work in mathematics, physics, philosophy, and religion. At age 14, he attended, with his father, the Paris meetings of
the circle of Marin Mersenne. He became familiar with the projective methods
of another member of Mersenne's circle, Girard Desargues, and applied them in
his only published work of mathematics, [77] of 1640. That work presented his
Hexagon Theorem, generally called Pascal's Theorem.

Charles-Julien Brianchon entered the *École Polytechnique* in 1804. An important
theorem, which he proved in 1806 [16], bears his name. Brianchon's proof of 1806
and that by Poncelet in 1822 made use of the *dual* of Brianchon's Theorem, which
was Pascal's Hexagon Theorem.

Here are the dual theorems, Pascal's Hexagon Theorem (See Fig. 6.4 Left) and
Brianchon's Hexagon Theorem (See Fig. 6.5.). Poncelet proved Pascal's Hexagon
Theorem by using the *Quatrième Principe*, or *Fourth Principle*, from 1813, which
we will see in the next section. The proof we present here is from [33, What is
Mathematics?]. Brianchon's Theorem follows.

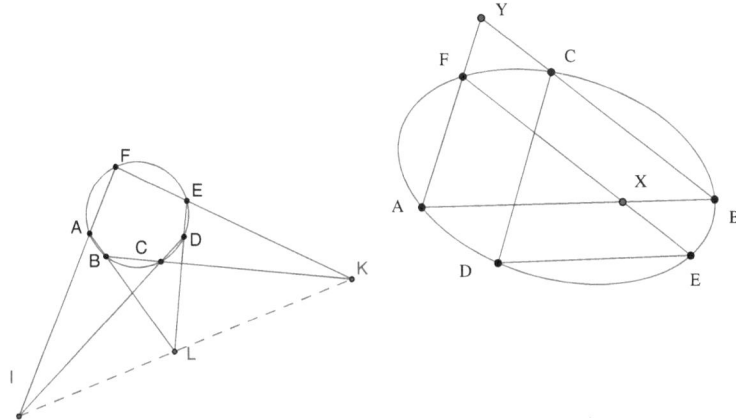

Fig. 6.4 Left: Pascal's Hexagon Theorem based on $Fig.$ 33 Poncelet's *Traité*. Right: Proof of Pascal's Hexagon Theorem by Courant and Robbins

Fig. 6.5 Brianchon's
Theorem, based on Poncelet's
$Fig.$ 33, in the *Traité* 1822

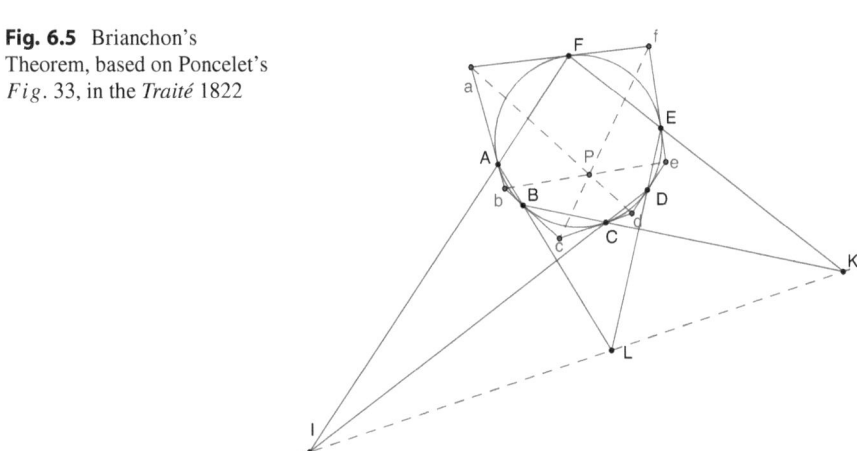

Theorem 6.3 (Pascal's Hexagon Theorem, 1640)   *If a hexagon ABCDEF is inscribed in a conic section, then the opposites sides meet in collinear points:* $I = AF \cap CD, \ L = AB \cap DE, \ K = BC \cap EF.$

Proof (Courant and Robbins) We are given hexagon $ABCDEF$ inscribed in a conic section. Let lines AF and CB meet in Y, and lines EF and AB meet in X. Let I and L be projected to infinity, giving the conic pictured in Fig. 6.4 Right, with $AB \parallel DE$ and $AF \parallel CD$. The cross-ratio of concurrent lines depends only on the angles at which the lines meet, and inscribed angles in a circle which intercept the same arc are congruent, so in a circle whose projection is the given conic, the cross-ratio of lines CF, CA, CB, CD on C equals the cross-ratio of lines EF, EA, EB, ED, respectively, on E. Projection preserves the cross-ratio, so the corresponding cross-ratios of lines meeting at C and at E on the conic are equal. The concurrent lines on C meet line FA in, respectively, F, A, Y, ∞, and the concurrent lines on E meet line AB in, respectively, X, A, B, ∞. So $CR(FA, Y\infty) = CR(XA, B\infty)$. This means

$$\frac{FY \cdot A\infty}{F\infty \cdot AY} = \frac{XB \cdot A\infty}{X\infty \cdot AB} \text{ i.e. } \frac{FY}{AY} = \frac{XB}{AB}.$$

This, in turn, means that lines CB and EF are parallel. So the points I, K, L all lie at infinity, so the original points are collinear.   □

In the 1806 paper proving Brianchon's Theorem, Brianchon used Pascal's Hexagon Theorem but credited Carnot's [21, Theorem 45] of 1803 instead of Pascal's 1640 work [77]. In [18] of January 1813, Brianchon correctly traced the hexagon theorem to Pascal's *Essai sur les Coniques*, of 1640.

Theorem 6.4 (Brianchon's Hexagon Theorem, 1806) *For a hexagon circumscribed about a conic section, the diagonals on opposite vertices are concurrent.*

In his 1822 *Traité*, Poncelet applied the pole-polar concept in a proof of Brianchon's Theorem [87, Art. 208]. Our proof follows the lines of Poncelet's.

Proof See Fig. 6.5. We assume the conic section is a circle. Let the circumscribed hexagon have sides tangent to the circle at A, B, C, D, E, F. Consider point L, the intersection of sides ED and AB of the inscribed hexagon. Tangents from E and D meet on the polar of L. Likewise, tangents at A and B meet on that same polar, so the dotted line eb is l, the polar of L. By Pascal's Hexagon Theorem, I, L, and K, the points at which the opposite sides of $ABCDEF$ meet, are collinear. Further, since I, L, and K are collinear, then by the Pole-Polar Theorem, the polars of I, L, and K are concurrent, at a point P. In other words, the three diagonals of the circumscribed hexagon are concurrent.

Any conic inscribed in a hexagon can be projected to a circle inscribed in a hexagon, preserving all the properties needed for the proof, so the theorem is proved for a hexagon circumscribed about a conic.   □

---

4   Poncelet's Fourth Principle and His Proof of the Pole-Polar Property

While in Russia, in 1813–1814, Poncelet discovered and proved his Fourth Principle, the *Quatrième Principe*, which greatly augmented the power of projective transformation. The set of principles concerning plane-to-plane projection was introduced in *Cahier* 3. We'll note Principles 1 and 2, and then present Principle 4 as a theorem.

The First Principle is that (*i*) every curve of second degree (i.e., conic section) is the intersection of a cone of circular base with a slicing plane, and (*ii*) such a curve is the *projection* of the circle of the base plane onto the slicing plane. Further, tangents to the conic section are projections of tangents to the base circle.

The Second Principle states that any set of parallel or concurrent lines have as their projection a set of parallel or concurrent lines.

Theorem 6.5 (Fourth Principle, Quatrième principe) *Any circle and line in the plane of the circle can be projected to another plane so the circle is projected to a circle and the projection of the line "passes to infinity." [84]*

(We assume that the line does not meet the circle. Poncelet does not make this restriction, just noting that when the line meets the circle then the circle produced will be imaginary.)

Poncelet's proof of the Fourth Principle makes use of the subcontrary circle, which we already encountered as Prop. 5 of Book 1 of the *Conics* of Apollonius. A simpler, and more limited, version of Poncelet's proof is in Appendix 4, at the end of this book.

The Fifth Principle extends the Fourth Principle to the case where the initial figure is a conic section. By the way a conic section is defined, it can always be projected to a circle in the base plane, and in that projection the given line is projected to a line in the base plane outside the circle. So the proof for a circle lets us reach the same conclusion.

To illustrate the power of the Fifth Principle, we look at Poncelet's 1813–1814 proof of an important theorem, which we have already proved as Monge's Pole-Polar Theorem. It pairs a line outside a conic and a point inside the conic. Recall that when the tangents from a point X outside a conic section meet the conic section at E and F, then line EF is the *polar* of X and X is the *pole* of EF.

Theorem 6.6 (Monge's Pole-Polar Theorem) *Given a conic and a line LM outside the conic, then there is a point, O, the* pole *of LM, inside the conic, which lies on the polars of all the points m of LM.*

Proof By Poncelet's Fifth Principle, there is a plane-to-plane projection so the conic is mapped to a circle and the line is mapped to the line at infinity. When the plane in this proposition is mapped so line LM goes to the line at infinity, then the tangents from a point such as m on LM are mapped to parallel lines. Our Fig. 6.6, based on Poncelet's *Fig.* 175, shows the figure before and after the projection. On the right, after projection, parallel tangents to a circle meet the circle in opposite points, points which are joined by a diameter. The diameters of a circle are concurrent at the center of the circle. Tangency is preserved by projection, so the center of the circle is paired with the point O that we seek.   □

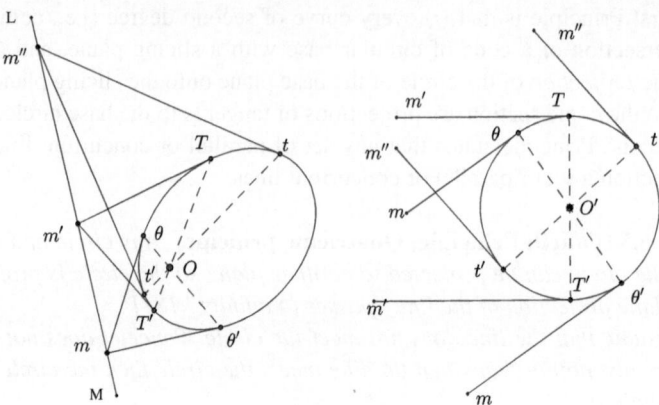

Fig. 6.6 Poncelet, *Cahier* 7 proof of Pole-Polar property, based on *Fig*. 175 1813/1862

5   Optional. Pole and Polar by George Salmon

This section violates our self-imposed rule to avoid coordinate or analytic arguments. However, this approach to geometry is a nineteenth century development that, by its elegance and simplicity, should be of interest to readers. This version is due to George Salmon (1819–1904), on the pole and polar pairing for a circle whose center is at the origin. Salmon was one of the most accomplished textbook writers of the nineteenth century. He was a student at Trinity College Dublin, where he became a tutor, and then a professor in both mathematics and divinity. His books, especially his *Treatise on Conic Sections* [91], are models of exposition, at the same time introducing new ideas in geometry to university students.

The *Treatise on Conic Sections* was first published in 1848. We follow the development of *pole* and *polar* from the third edition, of 1855. I will briefly use one tool not available to Salmon: two vectors $\vec{a}$ and $\vec{b}$ are perpendicular exactly when the dot product, $\vec{a} \cdot \vec{b}$, is 0. Also, in his efforts to simplify notation, Salmon represented the ordered pair (x, y) as, simply, xy. I will keep to the traditional (x, y). See pages 19 and 76–79 of [91].

First, what is the equation of the tangent line to circle $x^2 + y^2 = r^2$ at point (x'', y'') on the circle? Answer:

$$xx'' + yy'' = r^2.$$

Why? The tangent at (x'', y'') is perpendicular to the radius drawn to (x'', y''). So point (x, y) is on the tangent line from point (x'', y'') exactly when

$$(x - x'', y - y'') \cdot (x'', y'') = 0, \quad i.e., \ xx'' + yy'' - x''^2 - y''^2 = xx'' + yy'' - r^2 = 0.$$

Next, as Salmon wrote, "To find the equation of the line joining the points of contact of the tangents from any point (x', y') ." First, since (x', y') is on the tangent at (x'', y''), $x'x'' + y'y'' = r^2$. The equation of the line we seek is $x'x + y'y = r^2$ since, by its form, it is the equation of a line and is satisfied by both points (x'', y'') on the circle. (Poncelet had used this idea in his 1813 *Cahiers* [84, p. 250], but not in his works of 1820 and 1822.)

This means that the line $x'x + y'y = r^2$ is the *polar* of point (x', y'), and that (x', y') is the *pole* of line $x'x + y'y = r^2$ (with respect to circle $x^2 + y^2 = r^2$). This relationship still holds when there are no tangents from (x', y') to the circle, i.e., (x', y') is inside the circle.

We see that the key *Pole-Polar Theorem* holds: Line z, with equation $ax + by = r^2$, is the polar of point $Z = (a, b)$, and Z is on line $x'x + y'y = r^2$, i.e., $x'a + y'b = r^2$, exactly when the point (x', y') is on z, i.e., $ax' + by' = r^2$.

Salmon went on to point out that if (x', y') is on the circle, the *polar* is the tangent at (x', y').

Here is another example of George Salmon's abbreviated notation. [91, Art. 36, 113]. Corresponding to the equation of a line $Ax + By + C = 0$, let S denote the function of x, y: $Ax + By + C$; let S' denote another line $A'x + B'y + C'$. (We will let S be understood as the line $S = 0$. We assume the coefficients of x and y are not both 0.) If the two lines meet at point (x', y'), then (x', y') also makes the function $kS + jS'$ equal to 0 for any constants k and j. In other words, line $kS + jS'$, for k and j not both 0, is concurrent with lines S and S'.

Now, let us think of circles. This time, S and S' denote functions of form $x^2 + y^2 + Ax + By + C$, where we assume there is at least one real pair (x, y) that makes the function zero. Then $S - S'$ is a line. In fact, $S - S'$ is the common chord of circles S and S', since it is 0 where S and S' meet.

Now let S, S', and S'' be three circles. Their common chords, pairwise, are $S - S'$, $S' - S''$, and $S - S''$. Since $(S - S') + (S' - S'') = S - S''$, then the three lines are concurrent, for any point on two of the three lines must lie on the third. This is the Common Secant Theorem. The proof given here was original with Julius Plücker in 1827 [80].

---

6   Application 1. Brianchon's Conic Section Problem

This is the first of the problems in Brianchon's work of 1810, [17], of which Poncelet wrote in his 1822 *Traité*: "[I must recognize] that the first idea of my work I owe to reading that work." [p. *xxxiv*]

Theorem 6.7 (Brianchon 1810) *Suppose we are given a conic and n fixed collinear points, P_1, P_2, ..., P_n, none on the conic, and we seek $n - 1$ points on the conic: (x_1, y_1), (x_2, y_2), ..., (x_{n-1}, y_{n-1}), so the side on the first two such vertices, (x_1, y_1) and (x_2, y_2), lies on P_1, the side on the second and third vertices, (x_2, y_2) and (x_3, y_3), lies on P_2, and so on. We can select (x_1, y_1) as we like on the conic; then (x_2, y_2) is the other point on the conic and on the line on (x_1, y_1) and P_1. We continue this way until we have (x_{n-1}, y_{n-1}), on the conic. Then point (x_n, y_n) is* *determined, depending only on the choice of (x_1, y_1), and it ranges along a fixed conic that is independent of that choice of (x_1, y_1).*

Proof How do we find (x_n, y_n)? It must lie on the line joining (x_1, y_1) to P_n and on the line joining (x_{n-1}, y_{n-1}) to P_{n-1} and that is how we will define (x_n, y_n). The claim is that the point (x_n, y_n) lies on some fixed conic, sliding on that fixed conic as (x_1, y_1) slides on the given conic. Brianchon first makes a plane-to-plane projection, as described above, that maps the line on the fixed points to the line at infinity. As a result, in place of the n fixed points, we have slopes $m_1, m_2, m_3, ..., m_n$ of the n sides. This provides n linear equations in variables $x_1, y_1, x_2, y_2, ..., x_n, y_n$: $(y_{k+1} - y_k) = m_k(x_{k+1} - x_k)$ for $k = 1, 2, 3, ..., n - 1$, and $(y_1 - y_n) = m_n(x_1 - x_n)$. Further, we obtain $n - 2$ linear equations by substituting pairs (x_k, y_k) into the equation of the given conic and subtracting for consecutive values of k. If, for example, $n = 3$, $m_1 = 1$, $m_2 = 2$ and the given conic is $x^2 + 4y^2 = 4$, we have $x_1^2 + 4y_1^2 = 4$ and $x_2^2 + 4y_2^2 = 4$. Subtracting gives

$$(x_2^2 - x_1^2) + 4(y_2^2 - y_1^2) = 0, \text{ i.e., } 4(y_2 - y_1)(y_2 + y_1) = -(x_2 - x_1)(x_2 + x_1).$$

Now, $(y_2 - y_1) = m_1(x_2 - x_1)$, so

$$4m_1(y_2 + y_1) = -(x_2 + x_1).$$

In this way we get $n - 2$ additional linear equations. So we have a system of $2n - 2$ linear equations in $2n$ unknowns. In general, there will be a solution expressing (x_1, y_1) in the two parameters, x_n and y_n. Substituting into the equation of the given conic, $x^2 + 4y^2 = 4$, we get a second degree equation in x_n and y_n, as we claimed.

If we return to the original n collinear points, note that in a plane-to-plane projection, a second degree curve is mapped to a second degree curve. (Any line that twice meets a second degree curve in one plane corresponds to a line that twice meets the projected image in the other plane.)   □

7   Application 2. Poncelet's Application of a Theorem by Carnot

Poncelet, in his 1822 *Traité*, found a powerful application of an 1806 theorem of Lazare Carnot. This is one of many cases in which Poncelet was motivated to improve on work of Carnot. Carnot's theorem follows immediately from the power-of-points A, B and C with respect to a circle.

Theorem 6.8 (Carnot's Theorem, 1806) See Fig. 6.7 Left. *Let the three sides of a triangle ABC meet a circle, side AB meeting the circle in P and P', side AC meeting the circle in R and R', and side BC meeting the circle in Q and Q'. Then*

$$AP \cdot AP' \cdot BQ \cdot BQ' \cdot CR \cdot CR' = AR \cdot AR' \cdot BP \cdot BP' \cdot CQ \cdot CQ'.$$

Fig. 6.7 Left: Based on Carnot's *Fig* 8, 1806. Right: Based on Poncelet's Art. 39, 1822

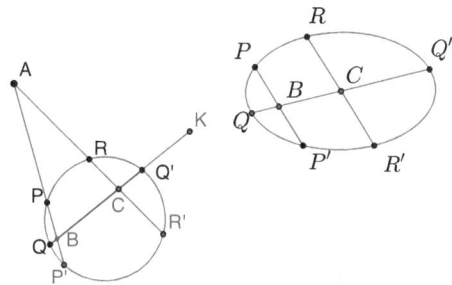

Theorem 6.9 (Poncelet, Art. 39, 1822) *Suppose that, in Carnot's Theorem, line BC is the polar of A and that K and C are harmonic conjugates of Q and Q'. Let the line AK be mapped to infinity. Then in the image, RR' and QQ' are conjugate diameters (diameters where each is an ordinate of the other) of conic section $QPRQ'R'P'$, while B is the midpoint of ordinate PP'. The equation of the conic section follows.*

Proof We leave unchanged the names of the points after the projection. See Fig. 6.7 Right. Recall that the midpoint of a segment MN is the harmonic conjugate of the point at infinity of line MN, with respect to M and N. So B and C are midpoints of chords PP' and RR' of the conic section that is the image of the circle; this means that chord QQ' is a diameter of the conic. Further, C is the midpoint of diameter QQ'.

We let $QC = CQ'$ be a and we let $RC = CR'$ be b. With these values and A at infinity, the equation of *Carnot's Theorem* above becomes

$$BQ \cdot BQ' \cdot b^2 = BP \cdot BP' \cdot a^2.$$

If we let BC be x (signed) and let PB be y, then the equation is $(a - x)(a + x)b^2 = y^2 a^2$, from which the standard equation of the ellipse follows.

(When line AK in the original figure meets the circle twice, we get a hyperbola.)

□

8   Exercises—Plane-to-Plane Projection

1. (Problem and solution by Louis Poinsot (1777–1859) in 1807 *Correspondance sur l'École Imperiale Polytechnique* [81, p. 306]) as reported by J. N. P. Hachette. The problem is now well known.
   Exercise Suppose we are given two lines, AB and CD, which meet off the paper we are dealing with, together with a point E on neither line. Construct by straightedge the line on E which is concurrent with AB and CD.

2. See Fig. 6.3, with the proof of Desargues' Theorem. Draw the diagram and prove Poncelet's claim that, after the projection, triangles BLM, DNP are similar and $LM \parallel NP$. In other words, given (*i*) $BL \parallel DN$, $BM \parallel DP$, and (*ii*) lines LN, BD, MP meet at H, then $\triangle BLM \sim \triangle DNP$ and $LM \parallel NP$.

3. Prove that any quadrilateral can be projected, in a plane-to-plane projection, to a parallelogram.

4. Prove the converse of Desargues' Theorem by projecting a line to the line at infinity. (In this case, the converse is also the dual.) In other words, *Given* triangles ABC and $A'B'C'$ where pairs of corresponding sides AB and $A'B'$, AC and $A'C'$, and BC and $B'C'$ meet, respectively, in collinear points L, M, N, *Show* that the triangles are in perspective from a point, i.e., lines AA', BB', CC' are concurrent. (The solution supposes line LMN is projected to the line at infinity.)

5. (i) On a piece of paper draw two lines, m and n, which are not parallel but will meet off the paper, and draw a point A on neither m nor n. By Desargues' Theorem, construct the line on A which is concurrent with m and n.

   (ii) On a piece of paper draw two parallel lines, m and n, and draw a point A on neither m nor n. By Desargues' Theorem, construct the line on A which is parallel to m and n.

6. (i) Let a quadrilateral $ABCD$ be inscribed in a conic section. Let opposite sides meet at U and in V. Show that U and V are collinear with the two points at which tangents to the conic at opposite vertices meet. (Note: A tangent at A is the limit of a secant AE as E approaches A on the path of the conic section.)

   (ii) Let a triangle ABC be inscribed in a conic section. Let the tangents at the three vertices meet their respective opposite sides of the triangle at L, M, N. Prove L, M, N are collinear.

7. [Poncelet 1822, Art. 546] Suppose that A, B, C, A', B', C' are on a conic section and $\triangle ABC$ and $\triangle A'B'C'$ are in perspective from a point P. Show that the line on which corresponding sides of the triangles meet is the polar of P with respect to the conic.

   Hint: See Construction L of Chap. 5.

8. [Poncelet 1822, Art. 203] Use Pascal's Hexagon Theorem. Given 5 points A, B, C, D, E of a conic and a line u on E, find the point F on u so that F lies on the same conic.

9. Let point A lie on a conic section. Prove that a line z is on A exactly when its pole, Z, lies on the polar of A.

10. Draw a circle and a point A outside the circle. By straightedge only, construct a tangent to the circle from A.

Homology as Developed by La Hire and Poncelet

7

1   Poncelet's *Homology*, and Homogeneous Coordinates

Starting with his *Cahiers* of 1813–1814, Poncelet developed the transformation that he called *homology*. As in Fig. 7.1, repeated from the last chapter, a simple form of homology is the projection of plane π to plane π' from an outside point, the center S. A point A of π is mapped to the point A' of plane π' that lies on line SA. The line at which the two planes meet is called the *axis*; it is formed of points mapped to themselves. The intersection of plane π and the plane on S that is parallel to plane π' is called the *vanishing line* since points on that line, such as X, have no finite image on plane π'. We say that the images of points of the vanishing line form the *line at infinity* of plane π'.

However, Poncelet allowed that a homology be carried out in one plane by superimposing the planes π and π'. One plane, say π', is rotated about the axis, until it coincides with the other plane, π.

When a homology operates in just one plane, what features appear in that plane? In the case of two planes, the *axis*, the line at which the planes meet, appears in the single plane as a *line of fixed points*; to have a homology operating in one plane, there must still be an axis. There will be a *line at infinity* and a *vanishing line*, where the vanishing line is mapped to the line at infinity. Except when they coincide, the axis, vanishing line, and line at infinity are parallel. (Problem: Name a transformation in which the three lines coincide.)

We summarize in a group of definitions.

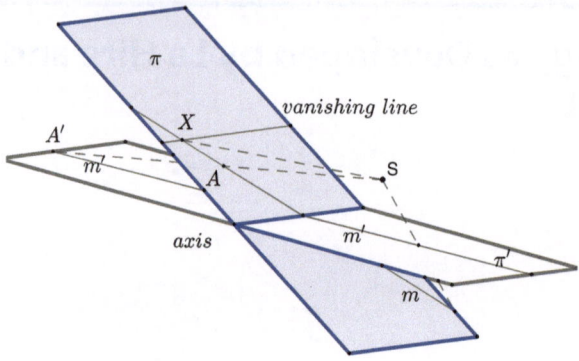

Fig. 7.1 Projection from point S of plane π plane π'

Definitions

The *Cartesian plane* or *real plane*, denoted $\mathbb{R}^2$, is the set of ordered pairs of real numbers, where the lines are the sets of points (x, y) that satisfy an equation $ax + by = c$ when a and b are not both zero.

The *real projective plane*, $\mathbb{P}^2$, is formed of the points and lines of the Cartesian plane together with *points at infinity*, one such point added to each line and shared with all parallel lines.

The set of points at infinity is itself a line, called the *line at infinity*.

A *collineation* is a one-to-one mapping of a plane onto itself in which lines are mapped to lines.

A *homology* is a collineation where there is a line of *fixed points*, called an *axis*. (A *fixed point* is a point mapped to itself.)

A *center* of a collineation is a point V such that every line m on V is mapped to itself, i.e., $m' = m$, but generally not every point of such a line m is mapped to itself.

In the projective plane, the line mapped to the *line at infinity* is called the *vanishing line*.

It is a theorem that a collineation with an axis must have a center.

In the twentieth century, as in H. S. M. Coxeter's [30, p. 247], a collineation with an axis is called a *perspective collineation*. An *elation* is a perspective collineation whose center lies on the axis, and other perspective collineations are called *homologies*.

We will later see how points at infinity can be added to plane $\mathbb{R}^2$ in a precise way.

2   Philippe de la Hire's *Plani-conique* of 1673

The first appearance of homology was in a short addendum, called *Plani-coniques*, to the 1673 *Nouvelle Methode* of Philippe de la Hire (1640–1718) [60]. The *Nouvelle Methode* was the first of La Hire's three general works on the conic sections. The second was *Nouveaux Elements des Sections Coniques . . .* [61], which developed the conic sections by coordinate geometry. The third was the widely known *Sectiones Conicae en novem libros distributae* [62] of 1685. It derived, by projective methods, the properties of conics found in Apollonius's *Conics*. Poncelet noted this 1685 work in the introduction to his 1822 *Traité*, but not the 1673 work.

In the *Plani-coniques*, La Hire's goal was to represent both the original figure and its homologue, or image, in a single plane,"without imagining any solid or plane except the plane of the figure." [60, p. 73]. In other words, thinking of a conic section as in Fig. 7.2, La Hire sought the rules of construction so that, point-by-point, a circle in a plane could be transformed to a conic section, the curve ELD in Fig. 7.2, but with the two planes superimposed by rotating one plane about the line at which the two original planes meet.

Since the base plane and the cutting plane in Fig. 7.2 meet in line FG, then FG is the *axis*, the line whose points are all mapped to themselves. La Hire found that the transformation is determined by a center, which he called the *pole*, on which all lines are mapped to themselves; an axis, called *formatrice*, a line of fixed points; and the line, the *directrice*, which we think of as the vanishing line, mapped to the line at infinity. In Fig. 7.2, line AK is parallel to line EG. So the line on K in the base plane that is parallel to FG is the *vanishing line*, or *directrice*, the line projected to infinity from point A onto the cutting plane.

K. G. C. von Staudt would show in 1847 that a collineation that has a *line of fixed points*, an *axis*, must also have a *center*, a point on which all lines are fixed. We will later see his proof. A different proof is by Dan Pedoe [79], in 1963, and his proof is developed in several Exercises at the end of this chapter. The dual holds: A collineation that has a center also has a line of fixed points.

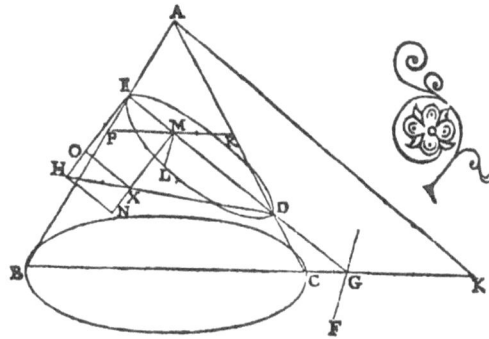

Fig. 7.2 Conic Section *ELD*, from the *Conics* of Apollonius, Commandinus 1696 edition. Conic section *EDL* is the projection of the base circle onto the slicing plane

Fig. 7.3 Construction of A'
in the homology with center
S and axis l, with X mapped
to X'

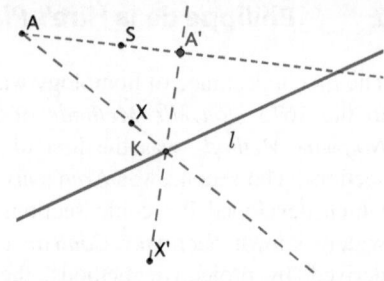

3   Poncelet's Construction of Images Under a Homology

As Poncelet would discover, the homology as defined by La Hire can also be
defined by designating the line of fixed points, l, which Poncelet called the *axis*,
by designating the *center*, S, and by pairing some other point X to its image X'
provided X, X', and S are collinear [87, Art. 302]. With this information, the
vanishing line can be found and it need not be given. Instead of giving two methods,
that of La Hire and that of Poncelet, for constructing image points in a homology, we
will only give that of Poncelet. La Hire's method will be developed in the Exercises.

Construction M Image of a point by a homology.

Figure 7.3 illustrates Poncelet's construction method. We are given the *axis*, l, the
center, S, and one point X and its image X', where S, X, and X' are collinear. We
will construct the image of a point A. We use the properties that lines are mapped to
lines, all points on the axis are fixed, and all lines on S are mapped to themselves.
We draw line AS, on which A' must lie. We will need to find the image of a second
line on A: draw line AX, which must meet l at a point—point K—and then draw
its image, line KX'. A' must be $KX' \cap AS$.

4   Application 1 of Homology: Inverse Homologues and
Composition of Homologies

In his Notebooks of 1813–1814, Poncelet used the term *homologues* for corresponding points and lines in systems of circles that are related by a similitude.
Already in the first Notebook, Poncelet observed another type of correspondence
of circles in a system, a fruitful concept that would be key in his development
of projective geometry. He would later call these pairs of corresponding points
inverse homologues, and this would be, in [85] and [87] of 1820 and 1822, his
prime example of a homology. (For details, see [9].) In Fig. 7.4, based on Poncelet's
Figures 8 and 10 of 1813/1862, T and t are inverse homologues, as are T' and t'. In
1813, Poncelet simply referred to T and t as "two exterior points," and to T' and t'
as "two interior points." The absence of special terminology in 1813 suggests that
Poncelet did not yet see the special role that these pairs of points would play.

Fig. 7.4 Poncelet
1813/1862, based on Figs. 7.8
and 7.10

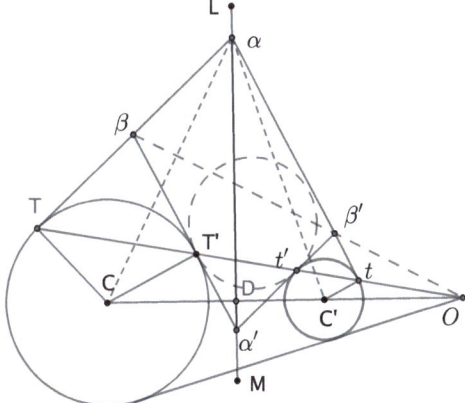

Definition
When two circles, C and C', are related by a dilation with center O, where t and T', in circles C and C', respectively, are homologues, if line OT' also meets circle C at T, then T and t are *inverse homologues*. And if line Ot also meets circle C' at t', then T' and t' are *inverse homologues*.

We shall prove, in the following section, that the pairing of inverse homologues of two circles is itself a homology. In other words, we will show that such a pairing is a collineation with a center and an axis. We will use the following theorem.

Theorem 7.1 *Let ϕ_1 and ϕ_2 be homologies with a common center S. Then the composition $\phi_1 \circ \phi_2$ is itself a homology with center S.*

Proof The composition of collineations is itself a collineation. It only remains to show that S is a center, i.e., S is collinear with any point X and its image $\phi_1 \circ \phi_2(X)$. Now, ϕ_2 is a homology so S, X, and $\phi_2(X)$ are collinear. And ϕ_1 is a homology so S, $\phi_2(X)$, and $\phi_1 \circ \phi_2(X)$ are collinear. So S, X, and $\phi_1 \circ \phi_2(X)$ are collinear. Any collineation with a center must be a homology. This completes the proof.   □

5   Application 2 of Homology: An Involution Induced by a Homology

We now look at the special case of a homology applied to a circle when the center of the homology, S, is a point outside the circle and the axis is the polar of S. We suppose the diameter on S meets the circle at A and B, and that the homology maps A to B. It turns out that the circle is mapped to itself. A corresponding theorem holds when S is inside the circle, but not the center of the circle.

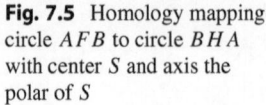

Fig. 7.5 Homology mapping
circle AFB to circle BHA
with center S and axis the
polar of S

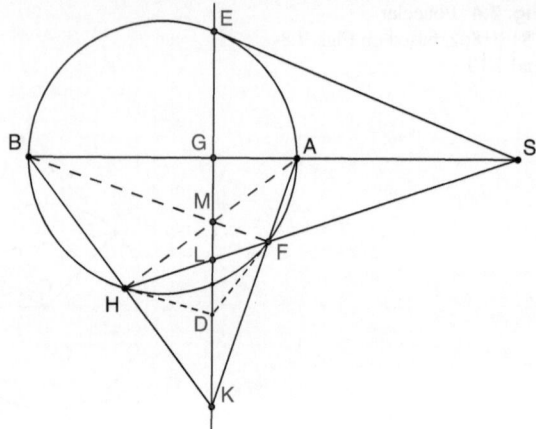

Theorem 7.2 (Circle-to-Circle Theorem) *Given a circle with diameter AB and point S on AB outside the circle, consider the homology, ϕ, with center S and whose axis is the polar of S, mapping A to B. Then*

(i) the homology maps B to A.
(ii) For any secant on S meeting the circle at H and F, the homology exchanges H and F. It follows that the circle is mapped to itself.

Proof See Fig. 7.5. We know that by specifying the *center*, S, the *axis*, EK, and that point A is mapped to point B, then the homology, ϕ, is defined. And we know $H(AB, SG)$, where G is the intersection of AB with the polar of S. By the definition of harmonic conjugates, $H(BA, SG)$. Since a homology maps a harmonic set to a harmonic set, then B must be mapped to A.

For *(ii)*, let a secant on the center, S, cut the circle at H and F and cut the polar of S at L. We know $H(HF, SL)$. Then, by the La Hire-Steiner Theorem, secants AF and BH meet on the polar of S, at K. Since the homology fixes K and interchanges A and B, then that homology interchanges lines BK and AK. So F and H must be interchanged.   □

Back in Fig. 7.5, fix point L on the polar of S, but let some point on line SL be F. Follow the same construction as before, letting AF meet polar EG at K, and then have KB meet line SL at H. As before, ϕ interchanges lines FK and HK so it interchanges the pair F and H.

We have just shown that the given homology, when restricted to line SL, is an *involution*. That collinear points can be related in an *involution*, and that involution is invariant under projection, was a crucial concept in Desargues' [35]. The concept was revived, with a modern definition by von Staudt [104, Art. 215] of 1847. Here is that definition.

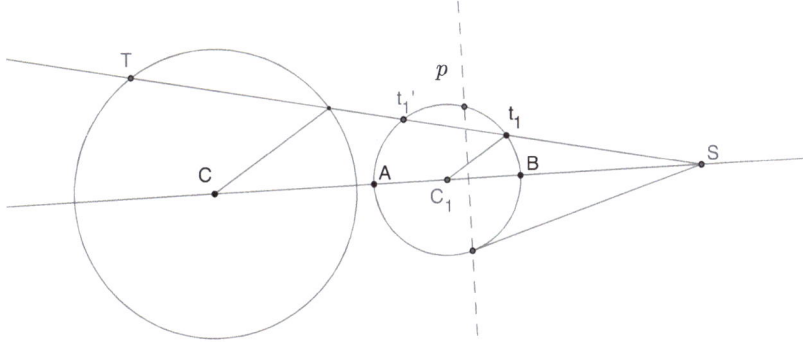

Fig. 7.6 Composition of involution on circle C_1 followed by similitude

Definition A function ϕ on a line is an *involution* if for every X on the line, $\phi(X)$ is mapped back to X. In other words, the composition of ϕ with itself is the identity function.

We will use the Circle-to-Circle Theorem to prove that the pairing of inverse homologues in two circles is a homology.

Theorem 7.3 *Given two circles which are not concentric, the pairing of inverse homologues is a homology.*

Proof See Fig. 7.6, where S is a center of the similitude, ϕ_2, mapping circle C_1 to circle C. (A similitude is a homology where the center of similitude is the center of the homology and the line at infinity is the axis.)

Let line p be the polar of point S with respect to circle C_1. Let ϕ_1 be the homology with center S mapping circle C_1 to itself, interchanging points A and B of the diameter which lies on S.

Then the composition $\phi_2 \circ \phi_1$ is a homology. It maps point t_1 of circle C_1 to t_1', and then maps t_1' to point T of circle C. In this way, inverse homologues are paired.

$\square$

6 Application 3 of Homology: Inscribed and Circumscribed Quadrilaterals to a Conic

Poncelet's *Traité* of 1822 is filled with conic properties that are proved by a projection. Art. 186 treats a quadrilateral inscribed in conic, with a related circumscribed quadrilateral. Figure 7.7 Left shows a conic with an inscribed quadrilateral, $ABCD$, whose opposite sides meet at E and at F, and the circumscribed quadrilateral whose

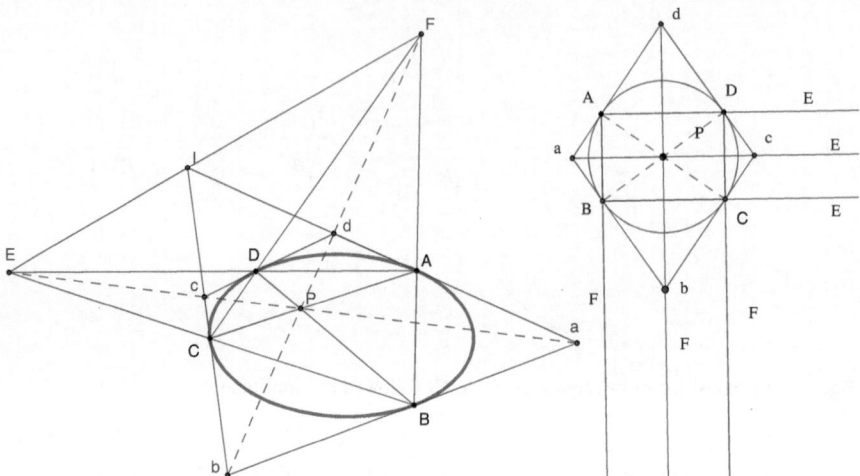

Fig. 7.7 Left: Conic, based on Poncelet's Art. 186, 1822. Right: Based on Poncelet's *Fig.* 27, 1822

sides are the tangents at A, B, C, D. Our Fig. 7.7 Right, based on Poncelet's *Fig.* 27, shows the figure after the conic is projected, with vanishing line EF, so the conic becomes a circle. Since $ABCD$ is inscribed in a circle and both pairs of opposite sides are parallel, then $ABCD$ must now be a rectangle. Further, the tangents drawn on the four vertices must form a rhombus, so its diagonals meet at the center of the circle, P. Conclude about the original conic

  (i) the diagonals of the inscribed and circumscribed quadrilaterals are concurrent,
 (ii) P, where the diagonals meet, is the pole of the line EF (providing an easy construction of the polar of a point inside a conic), and any point at which opposite sides of the circumscribed quadrilateral meet lies on EF.
(iii) The diagonals of the circumscribed quadrilateral lie on E and F.

7   Application 4 of Homology: A Homology Whose Center Is a Focus of a Conic

The concept of homology entered Poncelet's discussion of the focus, in a section devoted to pairs of conic sections.

Before we get into Poncelet's presentation, let us examine Poncelet's *Fig.* 69, in our Fig. 7.8. F and F' are the foci of the conic section and F is the center of the circle, c', on T'. The line on F and F' meets the conic at A and B. Let the given conic section be denoted c^*. Let the particular line $T'F$ meet the original conic c^* at T and T_1. (T_1 is not labeled on Poncelet's *Fig.* 69.)

Fig. 7.8 Based on Poncelet
1822, *Fig.* 69

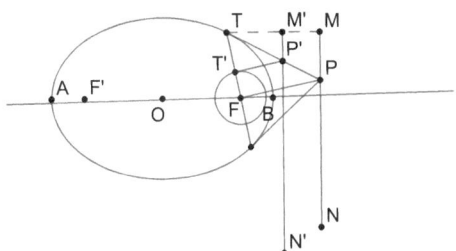

We know a circle will be mapped to a conic section by a homology, so let us examine this homology, ϕ, applied to the circle c' where the center of the homology is F and where that particular T' on c' is mapped to T, and the tangent to circle c' at T' is mapped to the tangent to conic c^* at T. Let the tangents at T and T' meet at P'. Therefore the axis of the homology is the line $M'N'$ on P'. Let P be on the polar of F with respect to the conic, where P is the intersection of the tangents to c^* at T and at T_1. By symmetry, $M'N'$ is perpendicular to AB, the major axis of the conic section. MN is drawn on P perpendicular to line AB.

Poncelet had reproved, in Art. 451, the property shown by La Hire in Book 8 Prop. 23 of [62], and which we state in the following lemma.

Lemma 7.1 (Poncelet 1822, Article 451) *Given a conic with focus F, and a line p on F. If P is the pole of p with respect to the conic, then line PF is perpendicular to p.*

Note that a homology maps a pole and polar with respect to a conic section to a pole and polar with respect to the image of the conic section. This is true since a tangent to a curve is mapped to a tangent of the image curve. The polar of F with respect to circle c' is the line at infinity.

By Lemma 7.1, line PF is perpendicular to chord TT_1. Tangent $T'P'$ is also perpendicular to TT_1 since it is a tangent to circle c'. This means that $FP \parallel T'P'$, so FP and $T'P'$ meet at infinity, at a point we denote as X. $\phi(X)$ lies on line FP. And $\phi(X)$ also lies on the image of line $T'P'$, which is line TP'. So $\phi(X) = P$. As we shall see, the same argument works for other points on MN, the polar of F with respect to the given conic, so ϕ maps the line at infinity to line MN.

Now assume we have drawn the image of circle c', which must be a conic section, and which we will name c. We wish to show that c is the given conic section. Return to Poncelet's $Fig.$ 69, but regarding chord TT_1 as any chord on F, meeting circle c' at T'.

We need to show that ϕ maps this new T' to T. Since line MN is the polar of F with respect to conic c^*, tangents to c^* at T and T_1 meet at P on MN, and line PF is perpendicular to chord TT_1. Again, the tangent to circle c' at T' is perpendicular to TT_1 and meets the axis $M'N'$ at the point P' as in $Fig.$ 69. Now, the tangent to circle c' at T' is mapped to the tangent to conic c at $\phi(T')$, and that tangent to c at $\phi(T')$ meets the tangent $T'P'$ at P' on axis $M'N'$. This means that $\phi(T')$ is T.

Fig. 7.9 Proof of Lemma,
base on Akopyan and
Zaslavsky

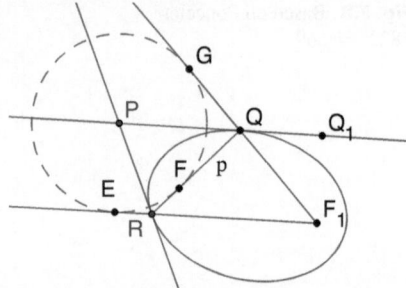

We conclude that conic c is the original conic c^*.

Before turning to the proof of Lemma 7, it is helpful to note various definitions of a conic section and its foci. Apollonius, followed by La Hire [62] and Poncelet, understood the conic section to be the intersection of a plane with a conic surface. The ensuing definition of the foci was complicated, and is only of historical interest today. As seen in a modern calculus book, a conic section is the solution set of second degree equation in x and y. (By this definition, a pair of lines is a conic section.) Foci are easy to define in terms of the lengths of the minor and major axes of a conic. Alternately, the conic itself can be defined starting with the foci: given a line l (*directrix*), a point F (*focus*) not on l, and a positive constant e (*eccentricity*), the set of points X so the ratio of distance FX to the distance from X to line l equals e. This focus-directrix characterization is found in Pappus's *Mathematical Collection*, Book 7 Prop. 238. That characterization of a conic section is assumed in the following proof of Lemma 7.1, while Apollonius and La Hire did not use the focus-directrix characterization of a conic section.

Proof (Akopyan and Zaslavsky, [3, p. 10]) See Fig. 7.9, the case of an ellipse. Let F and F_1 be the foci of the given conic section, where chord p, on F, meets the conic at Q and R. Let the tangents at Q and R meet at P, the pole of p. We define the conic by the property known to Apollonius: The ellipse with foci F and F_1 and major axis of length k is the set of points X such that $XF + XF_1 = k$. The angles of incidence formed by the tangent at Q are congruent: $\angle Q_1 Q F_1 \cong \angle PQF$, as are vertical angles $Q_1 Q F_1$ and PQG. Thus, QP is the bisector of angle GQF. We have corresponding congruences at R. So the circle with center P and tangent to p at some point Y is also tangent to lines $F_1 Q$ and $F_1 R$, at points G and E, respectively. Since $GQ = YQ$, $ER = YR$, and $F_1 G = F_1 E$, then $YQ + QF_1 = YR + RF_1$. By the definition of the ellipse, $FQ + QF_1 = FR + RF_1$. It follows that Y on p is F. Thus, the line PF is perpendicular to p.   □

For Poncelet, the focus-directrix characterization of a conic is a result of his analysis of $Fig.$ 69. Return to Fig. 7.8. Draw on T a parallel to the major axis, and assume M and M' lie on that line. Then since $T'P' \parallel FP$, $TF : T'F = TP : P'P = TM : M'M$. Since $T'F$ and $M'M$ are constants, then $TF : TM$ is a

Fig. 7.10 Proof of
concurrence at Gergonne
Point

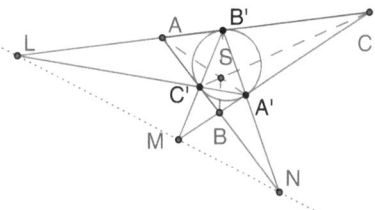

constant. Thus, Poncelet has derived the focus-directrix property of a conic, "très anciennement connue," "known from ancient times."

Michel Chasles, in [28, p. 180] of 1865, followed Poncelet in defining the focus, *foyer*, of a conic as a point S which serves as the center where the homologue of the conic is a circle. When Cremona produced, in 1893, an English version of his 1873 *Elementi di Geometria Projettiva*, he included a chapter on foci that was not in the 1873 original. There he defined foci as had Chasles [34, p. 249].

8 Application 5 of Homology: The Gergonne Point

We revisit the Gergonne Point, first treated in Exercise 19 of Chapter 1. Given a circle inscribed in triangle ABC, with AB tangent at C', BC tangent at A', and AC tangent at B'. Prove AA', BB', CC' are concurrent.

See Fig. 7.10. We define S to be $AA' \cap BB'$. We need to show CC' lies on S. Treat $A'A'B'B'C'C'$ as a hexagon inscribed in the circle. Side $A'A'$, for example, means the tangent to the circle at A', and its opposite side of the hexagon is $B'C'$. By Pascal's Hexagon Theorem, the three points where the tangents meet the opposite sides of the triangle, L, M, and N, are collinear. Apply the homology whose center is S and whose axis is LM, and which maps A to A'. Since B is on line NA, then the image of B is on line NA'. That image is also on BS, so B is mapped to B'. What is the image of C? C is on MB and on LA, so C is mapped to $MB' \cap LA'$. $MB' \cap LA'$ is C', so C, S, and C' are collinear. This is what we needed to show. □

9 Exercises—Homology

1. Explain why, under a homology, a line and its image must meet on the axis.
2. Explain why the axis of a homology and the vanishing lie can only meet at infinity, unless they are the same line. (We conclude that the axis and the vanishing line of a homology are parallel.)

Exercises 3–6 follow the proof of D. Pedoe in [79] of the following theorem:

Theorem 7.4 (Center-Axis Theorem) *Any collineation with an axis must have a center, and conversely.*

3. If a collineation has an axis, l, and a fixed point S which does not lie on l, then S is a center. Conversely, if a collineation has a center, S, and a fixed line l which does not lie on S, then l is a axis. (The converse is the dual).

4. Prove that a collineation that does not fix every point can have at most one center, i.e., at most one point S such that every line on S is mapped to itself.

5. Show: if a collineation has an axis, l, and a point X is not fixed, then line XX' is mapped to itself. Likewise, as the dual, if a collineation has a center, S, and a line m is not fixed, then point $m \cap m'$ is a fixed point.

6. Suppose a collineation has an axis, l, and no center on l, and P not a fixed point, then (*i*) there is a point Q not fixed so lines PP' and QQ' do not meet on l, and (*ii*) point $PP' \cap QQ'$ is a center.

   Note that the dual of a collineation with a line of fixed points is a collineation with a point on which every line is mapped to itself.

7. See Fig. 7.11 (*i*) Construct B', the image of B, under the homology with center S, axis as given, and which maps A to A'.

   (*ii*) In the same figure, find the point, C, on line SB that is mapped to infinity.

   In Exercises 8 and 9, you may draw parallels by approximating with a ruler.

8. and 9. Use the given Figs. 7.12 and 7.13 to find the images of the points E and D under the homology with center S, axis CB, and mapping A to A'. Then

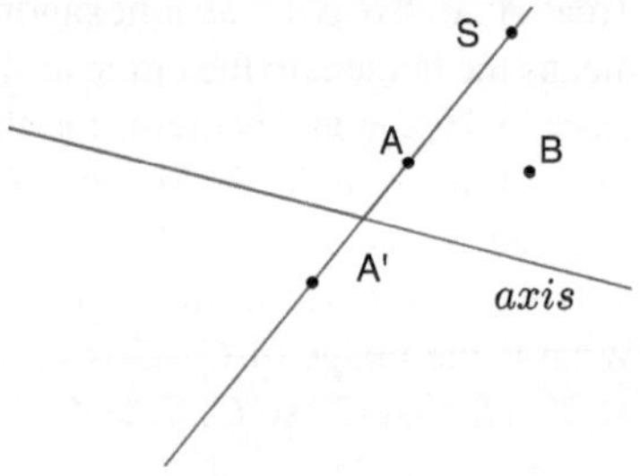

Fig. 7.11 Exercise 7. Find B' under the homology with center S and A mapped to A'

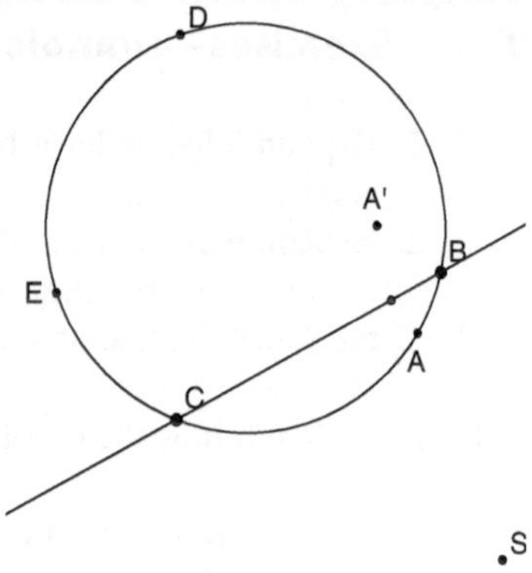

Fig. 7.12 Exercise 8. Homology with center S, axis CB, mapping A to A'

Fig. 7.13 Exercise 9. Homology with center S, axis CB, mapping A to A'

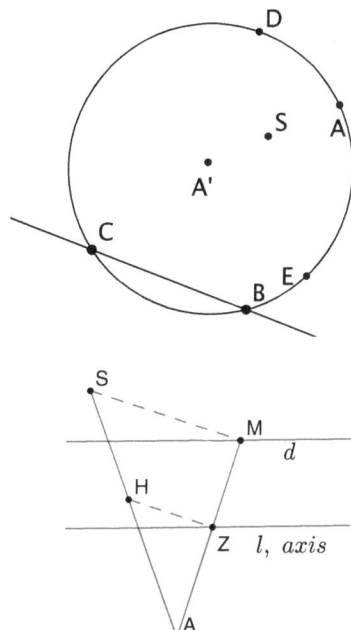

Fig. 7.14 Exercise 12. La Hire's 1673 construction of A' in the homology with center S, axis l, and vanishing line d

sketch the conic which is the image of the given circle. It may help to find the vanishing line or to find images of additional points of the circle.

10. Draw a line l to serve as axis, a circle c, a center S, and a point X and its image X' where S, X, and X' are collinear. Now construct images of points on circle c enough to see the shape of the conic section that is the image of c by the collineation defined this way.

11. Poncelet, after he presented his construction for points under a homology in Article 302 of 1822, offered several other ways to determine a homology.

    Here is a similar problem, the dual of the procedure of Art. 302. Given a center S and an axis l and a line m and its homologue m', where $m \cap m'$ is on l, find a point X and its homologue X' to define the homology which satisfies the given conditions.

12. (La Hire's construction [60, 1673]) As in Fig. 7.14, we are given the center, S of a homology, the axis, l, and the vanishing line, d. To find the image of a point A, we need to find two lines on which the image, A', is to lie. One line is AS. For a second line, we draw some other line on A, meeting the axis at Z and the vanishing line at M. To find the image of line AM, draw on Z a parallel to line SM, meeting line AS at H. Explain why H is A'.

13. Given a center V and three points, A, B, C, not collinear and no two collinear with V, and images A', B', C', respectively, not collinear and satisfying the condition that a point and its image are collinear with

V, prove there is a homology with center V and mapping A, B, C as indicated.

Hint. Apply Desargues' Theorem to triangles ABC and $A'B'C'$.

14. We know that a conic section is the image of a circle under a homology. And that the vanishing line of a homology is the set of points mapped to the line at infinity. The type of conic: parabola, ellipse, or hyperbola, depends on the relation of the vanishing line to the circle which is mapped to the conic.

Which conic do we get when (*i*) the vanishing line does not meet the circle, (*ii*) the vanishing line is tangent to the circle, and (*iii*) the vanishing line meets the circle twice? Why?

15. (a) Show that the composition of two homologies with the same axis must be a homology with the same axis. (Recall: A collineation with an axis must have a center.)

   (b) By (a), the composition of two given homologies, ϕ_1 and ϕ_2, with the same axis is a homology with that same axis. When we have the composition of two homologies with a common axis, show that the center of the resulting homology, $\phi_1 \circ \phi_2$, is collinear with the centers of the two given homologies.

Matrices and Homogeneous Coordinates

8

1   Homogeneous Coordinates

Moebius showed in [67], of 1827, how to precisely define a projective plane as formed of points that can be represented as triples of real numbers. He developed *homogeneous coordinates* based on physical considerations. He observed that for three fixed points A, B, C, not collinear, then for any point P in the plane of those points, there are weights, possibly negative, a at A, b at B, and c at C, so the center of mass of that three-body configuration is at P. In other words, $P = aA + bB + cC$. And if the triple (a, b, c) produces a center of mass at a point P, then so does (ka, kb, kc) for any non-zero k.

However, we, and nearly everyone else, follow the definition by Julius Plücker in 1831, which we simply refer to as *homogeneous coordinates*.

For the projective plane $\mathbb{P}^2$, we need to add points to the real plane $\mathbb{R}^2$ so that any collection of parallel lines will meet on one of these additional points, which we think of as lying at infinity. We represent each point (x, y) of $\mathbb{R}^2$ as $(x, y, 1)$ in $\mathbb{P}^2$. As we shall see, points at infinity will be represented by triples of the form $(x, y, 0)$ where x and y are not both zero.

Let us look at lines. Any line in $\mathbb{R}^2$ is the solution of an equation of the form $ax + by + c = 0$ where a and b are not both 0. A pair (x, y) is a solution of an equation $ax + by + c = 0$, exactly when $(x, y, 1)$ is a solution of $ax + by + cz = 0$. We will represent the line $ax + by + cz = 0$ by the triple $[a, b, c]$. Note that if $k \neq 0$, then equations $ax + by + cz = 0$ and $kax + kby + kcz = 0$ have exactly the same solutions. This means $[a, b, c]$ and $[ka, kb, kc]$ denote the same line; in this sense we have $[a, b, c] = [ak, bk, ck]$ in $\mathbb{P}^2$ whenever $k \neq 0$. This is the *homogeneity* property of lines in $\mathbb{P}^2$.

As Moebius noted, points have the *homogeneity* property. For $k \neq 0$, points (a, b, c) and (ka, kb, kc) lie on exactly the same lines. So in $\mathbb{P}^2$, triples (a, b, c) and (ka, kb, kc) represent the same point for $k \neq 0$. Any point (a, b, c) where $c \neq 0$

© The Author(s), under exclusive license to Springer Nature Switzerland AG 2025
C. Baltus, *Geometry by Its Transformations*, Compact Textbooks in Mathematics,
https://doi.org/10.1007/978-3-031-72281-3_8

can be put into a standard form $(x, y, 1)$. Triple $(0, 0, 0)$ is not a point of $\mathbb{P}^2$ since it would lie on every line.

What points at infinity do we add so that parallel lines will meet there? For the line with equation $ax + by + cz = 0$, with a and b not both 0, point $(-b, a, 0)$ lies on that line, and any triple $(x, y, 0)$ satisfying the equation must be a scalar multiple of $(-b, a, 0)$. On the other hand, which lines does point $(-b, a, 0)$ lie on? It lies on lines with equations $ax + by + cz = 0$, for those particular a and b, but for any c. Lines with equations $ax + by + cz = 0$ in homogenous coordinates correspond to the lines $ax + by + c = 0$ in the real plane for fixed values of a and b, as c ranges over the real numbers. That is a family of parallel lines in the real plane.

In $\mathbb{P}^2$, all points of form $(a, b, 0)$, where a and b are not both 0, lie on the new line $z = 0$, or $[0, 0, 1]$, the *line at infinity*.

We summarize.

> **Definition**   The *real projective plane*, $\mathbb{P}^2$, is the set of triples (x, y, z) for any real numbers x, y, z not all 0. If $k \neq 0$, then triples (a, b, c) and (ka, kb, kc) represent the same point. The lines are triples $[a, b, c]$ where a, b, c are not all 0, and for $k \neq 0$, $[a, b, c]$ and $[ka, kb, kc]$ represent the same line. A point (x, y, z) lies on line $[a, b, c]$ exactly when $ax + by + cz = 0$.

2   Collineation Defined by Matrix Multiplication

A *collineation* is simply a one-to-one and onto mapping of an affine or a projective plane to itself in which collinear points are always mapped to collinear points. A major theorem of Moebius is that there is a unique collineation mapping any four given points of the projective plane, no three collinear, to any four given points, no three collinear. He described the collineation in terms of equations. We will examine it in essentially the same way, but represented by a three-by-three matrix.

All the transformations we will discuss, with the exception of the inversion transformation, can be carried out by matrix multiplication. Matrix representations offer insight into the transformations themselves and provide a way, easily employing computer operations, to carry out computations.

For this book, a *vector* is an ordered pair or triple of real numbers. We will work in $\mathbb{R}^2$, the real plane, formed of real ordered pairs, or in $\mathbb{P}^2$, the projective plane, formed of ordered triples of real numbers (a, b, c), where a, b, c are not all 0. (We will also work, occasionally, in $\mathbb{R}^3$, where, for example, $(1, 2, 3)$ and $(2, 4, 6)$ are different points. Care is needed.)

For $\mathbb{R}^2$, the basis will be the *standard basis* $\{\vec{e_1}, \vec{e_2}\}$ where $\vec{e_1} = (1, 0)$ and $\vec{e_2} = (0, 1)$. This means, for one thing, that any vector (a, b) is $a\vec{e_1} + b\vec{e_2}$. In $\mathbb{R}^3$, $\vec{e_1} = (1, 0, 0)$, $\vec{e_2} = (0, 1, 0)$, and $\vec{e_3} = (0, 0, 1)$. ($\mathbb{P}^2$ is not a vector space. It has, for example, no identity element for addition.)

Readers unfamiliar with matrices and matrix algebra will find an introduction in Appendix 2 at the end of this book.

Matrices can execute transformations which are *linear*.

Definition
When $\vec{a}$ and $\vec{b}$ are *vectors* and α and β are *scalars*—which for us are real or complex numbers—then a transformation f is *linear* if $f(\alpha\vec{a} + \beta\vec{b}) = \alpha f(\vec{a}) + \beta f(\vec{b})$. When it is clear from the context that a vector is intended, the arrow will be omitted in the vector symbol.

Matrix multiplication is a linear transformation in that when M is a matrix and $\vec{x}$ and $\vec{y}$ column vectors of the correct dimension, then $M(\alpha\vec{x} + \beta\vec{y}) = \alpha M\vec{x} + \beta M\vec{y}$.

By the way matrix multiplication works, if, say, a 3-by-3 matrix M is $[\vec{c}_1 \ \vec{c}_2 \ \vec{c}_3]$ where $\vec{c}_1$ is the first column, etc., then we have:

the matrix product $M\vec{e}_1$ is the column vector $\vec{c}_1$, when $\vec{e}_1$ and $\vec{c}_1$ are written as column vectors, and so on. We have a corresponding rule for 2-by-2 matrices.

Example 8.1 Rotation about the origin by angle θ in the positive direction sends $\vec{e}_1$ to $(cos\ \theta, sin\ \theta)$ and sends $\vec{e}_2$ to $(-sin\ \theta, cos\ \theta)$, so a vector $\begin{bmatrix} x \\ y \end{bmatrix}$ is mapped to

$$\begin{bmatrix} cos\ \theta & -sin\ \theta \\ sin\ \theta & cos\ \theta \end{bmatrix} \begin{bmatrix} x \\ y \end{bmatrix}.$$

Example 8.2 A *dilation*, with center at the origin and scale factor k, maps (x, y) to (kx, ky) where we assume $k \neq 0$. In $\mathbb{R}^2$ and in $\mathbb{P}^2$, respectively, this transformation is effected by matrices $\begin{bmatrix} k & 0 \\ 0 & k \end{bmatrix}$ and $\begin{bmatrix} k & 0 & 0 \\ 0 & k & 0 \\ 0 & 0 & 1 \end{bmatrix}$.

Example 8.3 *Translation* by (r, s) in $\mathbb{R}^2$ is not linear, but the corresponding translation in $\mathbb{P}^2$ is linear, carried out by matrix $T = \begin{bmatrix} 1 & 0 & r \\ 0 & 1 & s \\ 0 & 0 & 1 \end{bmatrix}$. We see that this matrix multiplication maps $(x, y, 1)$ to $(x + r, y + s, 1)$ and $(x, y, 0)$ to $(x, y, 0)$.

This last note tells us that translation maps a line to a parallel line, since a given line and its image will meet on the line at infinity. Also note that the *inverse* of an operation f undoes that operation. In other words, if f maps a vector $\vec{x}$ to $f(\vec{x})$ then the inverse operation maps $f(\vec{x})$ to $\vec{x}$. In the following example, we will employ the inverse operation to translation.

Example 8.4 If we wish to carry out a rotation by θ about point $(r, s, 1)$, we can first translate so point $(r, s, 1)$ is mapped to the origin, $(0, 0, 1)$, by matrix T^{-1} (inverse matrix), then rotate by θ about the origin, and then, by matrix T, apply the inverse translation. The matrix for rotation by θ about the origin in $\mathbb{P}^2$ will be

$$R = \begin{bmatrix} cos\ \theta & -sin\ \theta & 0 \\ sin\ \theta & cos\ \theta & 0 \\ 0 & 0 & 1 \end{bmatrix}.$$ The composition of linear transformations is accom-

plished by the matrix product. So the desired rotation of point $(x, y, 1)$ about $(r, s, 1)$, is carried out by the matrix product $T R T^{-1}$:

$$\begin{bmatrix} 1 & 0 & r \\ 0 & 1 & s \\ 0 & 0 & 1 \end{bmatrix} \begin{bmatrix} cos\ \theta & -sin\ \theta & 0 \\ sin\ \theta & cos\ \theta & 0 \\ 0 & 0 & 1 \end{bmatrix} \begin{bmatrix} 1 & 0 & -r \\ 0 & 1 & -s \\ 0 & 0 & 1 \end{bmatrix} \begin{bmatrix} x \\ y \\ 1 \end{bmatrix}.$$

Theorem 8.1 *Any transformation carried out by matrix multiplication maps a line to a line.*

Proof Matrix multiplication is a linear transformation. In both $\mathbb{R}^2$ and $\mathbb{P}^2$, the line on points $\vec{a}$ and $\vec{b}$ is the set of points of the form $\vec{a} + t(\vec{b} - \vec{a})$ where t is any real number. The formula tells us that these are the points we reach by starting at $\vec{a}$ and then moving in the direction of vector $\vec{b} - \vec{a}$, or its opposite. This form is most commonly written as $(1 - t)\vec{a} + t\vec{b}$. If we fix t, and, therefore, fix some point on the line on $\vec{a}$ and $\vec{b}$, and multiply by matrix M, we get $(1 - t)M\vec{a} + tM\vec{b}$, a point on the line on points $M\vec{a}$ and $M\vec{b}$. Thus a line is mapped to a line.   □

We can conclude that any transformation defined by multiplication by a 2-by-2 or 3-by-3 matrix M is a collineation, a function that maps any line to a line. One needs to be careful because a transformation must be one-to-one and onto. And this brings up the concept of the *multiplicative inverse*, or just *inverse*, of a square matrix and the *determinant* of a matrix.

We note that a square matrix has an inverse matrix exactly when its determinant is non-zero. Here we look only at the case of the 2-by-2 matrix. There are more complicated rules for determinants and inverses for larger square matrices, which can be found in any linear algebra text.

Definition

The number $ad - bc$ is the *determinant* of matrix $M = \begin{bmatrix} a & b \\ c & d \end{bmatrix}$. The square

matrix $I = \begin{bmatrix} 1 & 0 \\ 0 & 1 \end{bmatrix}$ is the *identity matrix* since for any 2-by-2 matrix M, $MI = IM = M$.

(continued)

When the determinant $ad - bc$ of M is not zero, then the matrix $\frac{1}{ad - bc}\begin{bmatrix} d & -b \\ -c & a \end{bmatrix}$ is M^{-1}, the *inverse* of matrix M, in that $MM^{-1} = M^{-1}M = I$. Those square matrices which have inverse matrices are called *invertible*.

Note that the determinant $ad - bc$ of a 2-by-2 matrix is 0 exactly when one row or column is made of zeros or one row (or column) is a scalar multiple of the other row (or column). Without proof—which, again, can be found in a linear algebra textbook—here is an important theorem about determinants, and an important consequence in a second theorem.

Theorem 8.2 *The determinant of a square matrix M is 0 exactly when M is not invertible, which is the case exactly when one column (or row) of M can be written as the sum of scalar multiples of the other columns (or rows).*

Theorem 8.3 *When M is an invertible square matrix, then for any vector b of appropriate size, the equation $Mx = b$ has exactly one solution, x.*

Proof The inverse matrix M^{-1} exists since M is invertible. Multiply both Mx and b on the left by M^{-1}. This gives $x = M^{-1}b$, the unique solution. □

Theorem 8.4 *Three points of $\mathbb{P}^2$ are collinear exactly when the determinant of matrix M is zero, where the three points are the columns of M.*

Proof In the projective plane $\mathbb{P}^2$, the line on points A and B is the set of points of form $(1 - t)A + tB$ for real numbers t. A point C is collinear with A and B exactly when it has the form $(1 - t)A + tB$ for some real number t. Note that if $C = vA + wB$ for real v and w, then by the homogeneity property, C can be expressed in form $(1 - t)A + tB$. So A, B, and C are collinear exactly when C can be written as the sum of scalar multiples of A and B, i.e., the determinant of M is 0. □

3   Matrices and the Affinity Transformation

Let us consider an affinity. Moebius first introduced that transformation in terms of a triangle with vertices A, B, C, i.e., three points that are not collinear. Now any point P can be written in the form $aA + bB + cC$, where a, b, c are scalars, not all zero. To see this, let A, B, C, and P be written as triples in homogeneous coordinates; then (a, b, c) is the solution of the equation $M\vec{x} = \vec{P}$ where the columns of M are vectors A, B, C, and P is a column vector. Since A, B, and C are not collinear, then $Det(M) \neq 0$, so there is a unique solution (a, b, c) of the matrix equation. Moebius went on to say that an affinity is defined by designating three (finite) points A', B', C', not collinear, as the images, respectively, of A, B, C, and defining P', the image of P, to be $aA' + bB' + cC'$. In $\mathbb{P}^2$, $aA' + bB' + cC'$ equals $M'\vec{d}$ where matrix M' is formed of columns A', B', C', and $\vec{d}$ is the column vector form of (a, b, c).

Theorem 8.5 *Let a transformation on the projective plane be defined in the manner prescribed by Moebius: non-collinear finite points A, B, C are mapped to non-collinear finite points A', B', C', and any point $P = aA + bB + cC$ for scalars a, b, c is mapped to $P' = aA' + bB' + cC'$. Then the transformation is an* affinity, *in that parallel lines are mapped to parallel lines.*

Further, since a transformation carried out by matrix multiplication is linear, then the transformation has the property specified by Moebius: when a point X divides a segment AB in a certain ratio, then X' divides the image segment $A'B'$ in the same ratio.

Proof Thinking in terms of homogeneous coordinates, a collineation in $\mathbb{P}^2$ maps parallel lines to parallel lines—the definition of an affinity—exactly when points at infinity are mapped to points at infinity. (Parallel lines in $\mathbb{R}^2$ meet at a point at infinity in $\mathbb{P}^2$ and lines which meet at infinity in $\mathbb{P}^2$ are parallel in $\mathbb{R}^2$.) The points A, B, C, which are vertices of a triangle, are in $\mathbb{R}^2$ so the matrix M can be written with its bottom row $(1, 1, 1)$. If a point P is at infinity, its vector is of form $(r, s, 0)$. So when P is written in form $aA + bB + cC$ for scalars a, b, c not all zero, we see $a + b + c = 0$. So when the image, P', of P, is expressed in form $aA' + bB' + cC'$, where vectors A', B', C' are each written in form $(t, u, 1)$, then P' has form $(v, w, 0)$. Thus, the mapping defined by Moebius is, indeed, an affinity.

The second part follows from the fact that matrix multiplication is linear: point $X = tA + (1 + t)B$ is mapped to $tA' + (1 + t)B'$.   □

---

4   Matrices and the Collineation Transformation

The final transformation that Moebius introduced in 1827 was the *collineation* of the projective plane. As with an affinity, lines are mapped to lines, but there is no further requirement. Although Moebius did not use this terminology, a collineation of the projective plane is a *projectivity*. The name *projectivity* was used by Jacob Steiner in his book [98] of 1832, possibly for the first time.

There is a remarkable property, proved by Moebius.

Theorem 8.6 *Let the four points A, B, C, P have no three collinear, with intended images A', B', C', P', also with no three collinear. Then those four points and their images uniquely define a collineation on $\mathbb{P}^2$.*

Proof We will define the matrix M that effects the collineation. As with an affinity, there are scalars a_1, b_1, c_1 so $P = a_1A + b_1B + c_1C$, and scalars a_2, b_2, c_2 so

$P' = a_2 A' + b_2 B' + c_2 C'$. None of the scalars is 0 since no three of the four points are collinear.

We can find matrix M so

$$MA = \frac{a_2}{a_1} A', \quad MB = \frac{b_2}{b_1} B', \quad MC = \frac{c_2}{c_1} C'.$$

To see this, let Q be the matrix whose columns are A, B, C, and let Q' be the matrix whose columns are $\frac{a_2}{a_1} A'$, $\frac{b_2}{b_1} B'$, $\frac{c_2}{c_1} C'$. We need to find matrix M so $MQ = Q'$. There is matrix Q^{-1} since the determinant of Q is not 0. So $M = Q' Q^{-1}$.

Then

$$MP = M[a_1 A + b_1 B + c_1 C] = a_1 MA + b_1 MB + c_1 MC$$

$$= a_1 \frac{a_2}{a_1} A' + b_1 \frac{b_2}{b_1} B' + c_1 \frac{c_2}{c_1} C' = P'.$$

[67, Part 2 Art. 220].   □

---

5   Exercises—Homogeneous Coordinates

1. Write in form $Ax + By + Cz = 0$, in homogeneous coordinates, the equation of the line on points $(2, -4, 2)$ and $(2, 1, -2)$.
2. On line $3x - y + 2z = 0$, what is the point at infinity?
3. What is the line on point $(2, 1, -2)$ which meets line $x + 3y + 2z = 0$ at infinity?
4. Suppose the determinant $ad - bc$ of matrix $M = \begin{bmatrix} a & b \\ c & d \end{bmatrix}$ is not zero. Show by multiplication that the matrix $\frac{1}{ad - bc} \begin{bmatrix} d & -b \\ -c & a \end{bmatrix}$ is M^{-1} in that $MM^{-1} = M^{-1}M = I$.

Projective Geometry: Steiner and von Staudt

9

1 Steiner, 1832, and von Staudt, 1847, and Projective Geometry

Poncelet's *Traité* of 1822 bears little resemblance to the treatises on projective geometry that would follow. The 1822 work seems closer to what Lazare Carnot wrote in the first decade of the nineteenth century.

The first that resembles modern work is Jacob Steiner's *Systematische Entwicklung der Abhängigkeit geometrischer Gestalten, Systematic Development of the Relationship of Geometric Figures*, of 1832 [98]. The next major development came with the 1847 *Geometrie der Lage, Geometry of Position*, of Karl Georg Christian von Staudt. Von Staudt wrote right in the introduction to his [104], of 1847, that there is a "geometry of position" distinction from metric geometry, where, presumably, all that came before was "metric geometry." His goal was to make geometry of position an independent science which does not depend on length or angle measure, features of a figure which are not preserved by projection. Many geometers since the time of von Staudt have felt that a true projective geometry cannot depend on length and angle measure.

We will give an overview of the achievement of Steiner and von Staudt, and then present their ideas in sequence.

The *projective transformation* had been introduced by Moebius in his [67] of 1827. He had called it a collineation, to be applied in the projective plane. For Poncelet, *projection* almost always meant a plane-to-plane projection and the composition of projections appeared only incidentally in his 1822 work. With Jacob Steiner's *Systematische Entwicklung*, attention moved from plane-to-plane projection to mapping one elementary form to another elementary form, and to the composition of those mappings.

What essential characteristics did Steiner and von Staudt consider?

C. Baltus, *Geometry by Its Transformations*, Compact Textbooks in Mathematics, https://doi.org/10.1007/978-3-031-72281-3_9

(*i*) There are *elementary transformations* which relate *elementary forms*, and these
elementary transformations are combined by function composition to build the
full collection of *projective transformations*.

(*ii*) For a given pairing of elements of one geometric form with elements of a
second form, Steiner and von Staudt both asked which property distinguishes
the pairing as *projective*? This was not a question that Poncelet addressed.
Steiner's answer involved the cross-ratio, while von Staudt's answer involved
harmonic sets.

(*iii*) Duality is a necessary property of the system, present from the very beginning
in describing the geometric forms to be considered and the transformations that
projectively pair the elements of two forms. Poncelet had suggested duality in
only a few of his propositions. He said nothing, for example, on the duality of
Pascal's Hexagon Theorem and Brianchon's Hexagon Theorem.

(*iv*) The geometry developed should effectively include the propositions of a
projective nature in the tradition going back to Apollonius and up to the recent
geometry of Carnot and Brianchon. Steiner and von Staudt would situate these
propositions on their new framework of projective geometry.

Jacob Steiner (1796–1863) grew up on a farm in Switzerland. He didn't begin
formal schooling until age 18. The school he attended was the Pestalozzi school,
noted for stressing independent thinking, a characteristic of Steiner's professional
life. After that school, he supported himself by tutoring mathematics and then
attended lectures at the University of Berlin from 1822 to 1824. In those years,
his career as a prolific writer of mathematics was already underway.

A long multi-part paper by Steiner, on geometry, appeared in 1826, in the first
issue of *Crelle's Journals* [96]. This paper brought Steiner to the attention of French
mathematicians, in particular Gergonne, the editor of the French *Annales*. Steiner's
Systematische Entwicklung appeared in 1832. In recognition of the high quality of
his work, Steiner was appointed professor of geometry at the University of Berlin in
1834, a post created for him and which he held until his death.

In his 1826 work, Steiner introduced the term *power-of-a-point*, which we have
already seen. In another article in the same 1826 issue of *Crelle's Journal*, Steiner
showed that for two circles, the points of equal powers with respect to those two
circle are exactly the points of a line, which he called the *line of equal powers* and
which, unknown to Steiner, Poncelet had called the *common secant*.

Karl Georg Christian von Staudt (1798–1867) attended the University of Göttingen for three years, beginning in 1819. There von Staudt attended lectures of
Gauss, who was a professor and ran the observatory. Von Staudt then moved back
nearer to home, becoming a mathematics teacher at the gymnasium level, while
also engaged in mathematics research. In 1835, he was appointed professor of
mathematics at Erlanger, a position he held for the rest of his life.

2   The Projective Geometry of Steiner

We begin our step-by-step development, as Steiner did, with his list of the *elementary forms*.

1. **Elementary Forms** Steiner noted the basic ideas of geometry: space, plane, line and point. Then he introduced the five *elementary forms*, *Grundebilder*.
    (*i*) A *line* of points, which includes a point at infinity,
    (*ii*) A *pencil* of concurrent lines in one plane. *Pencil* is modern vocabulary, denoting the set of all lines of a plane on a point. Steiner used the term *ebener Strahlenbüschel*, a tuft of rays in a plane. The point on which the lines lie is the *center* of the pencil. A pencil is the dual of the set of points on a particular line.
    (*iii*) The set of all the planes, forming an *Ebenenbüschel*, on a particular line in space.
    (*iv*) The plane.
    (*v*) The set of all lines in space on a particular point.
    We will not deal with (*iii*) and (*v*), but it is important to note that from the time of Steiner, three-dimensional forms were part of most projective works.
2. **Fundamental Relations** Steiner then moved to the "fundamental relations" among the elementary forms. He distinguished between a *perspectivity* and a *projectivity*.

Definitions
A *perspectivity* pairs two elementary forms in a single projection. We pair a pencil of lines B with another pencil B_1 by specifying a line m on neither B not B_1, and then pairing each line on B with the line on B_1 that it meets on m. As the dual, we pair the points of two lines, b and b_1, by means of a point M on neither line, where each point X of b is paired with the point X' of b_1 that lies on line MX. See Fig. 9.1. Finally, we can pair a pencil, with center M, and a line, l, not on the center of the pencil, in the obvious way: a line m_1 of the pencil is paired with the point $l \cap m_1$.

Then a *projectivity* is the *composition* of one or more perspectivities, and two forms related by a perspectivity or a projectivity are *projectively related*.

---

3. **Projectively Related Forms** An important property is that the cross-ratio involving collinear points or concurrent lines is preserved by any perspectivity. Therefore, the cross-ratio is preserved by any projectivity. Further, we have seen that for any three given collinear points A, B, C, a fourth collinear point D is determined by the value of the cross-ratio. This is because there is exactly one point D which makes $CR(AD, BC) = k$ for any k except 0, 1, or infinity. The dual claim must hold, too [98, p. 252]. This theorem follows.

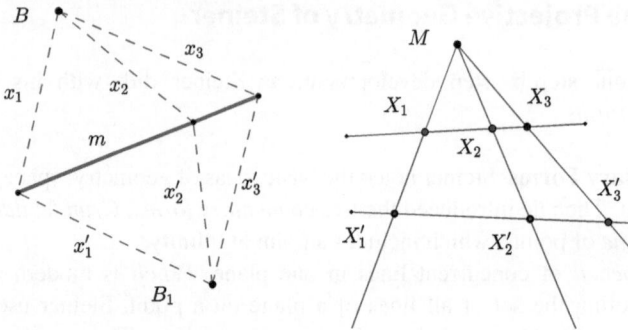

Fig. 9.1 Perspectivities as by Steiner

Theorem 9.1 (Steiner)

(i) *Two elementary forms are* projectively related *exactly when the cross-ratio of any four elements of one form equals the cross-ratio of the corresponding four elements of the other form.*

(ii) *A projective relationship between points of two lines is completely determined by pairing three distinct elements of one line with three distinct elements of the second line. More generally, the same holds for any two forms that are projectively related. This is the Fundamental Theorem of Projective Geometry.*

4. **Duality** Steiner emphasized *duality*. Many of Steiner's propositions appeared in dual pairs, in two columns on the page, as in Gergonne's [49] of 1827. Paired as duals we have the *complete quadrilateral*, consisting of four lines, with no three concurrent, and the six points at which pairs of those lines meet, and the *complete quadrangle*, consisting of four points, with no three collinear, and the six lines joining pairs of those points.

We have seen that the cross-ratio of four collinear points has its dual. For both, Steiner used the name *Doppelverhältniss*. As we have seen, the cross-ratio of collinear points A, B, C, D, which we denote as $CR(AB, CD)$, is $(AC \cdot BD)/(AD \cdot BC)$. Nowadays, these are directed lengths, which can be negative, but Steiner took them as positive lengths, where the order of the four points was to be specified. Steiner's proof of the invariance of the cross-ratio under projection follows the strategy we have seen, used by Carnot in 1806 for invariance of the harmonic relation.

5. **Composition of Transformations and Desargues' Theorem** Steiner employed the composition of transformations, something Poncelet did not do except in rare cases.

Here is an important proposition [Art. 21], from which Desargues' Theorem follows immediately. See Fig. 9.2. Let lines A, A_1 and A_2 meet in point e. Project line A to line A_1 from a point B: a to a_1, c to c_1, and e to itself.

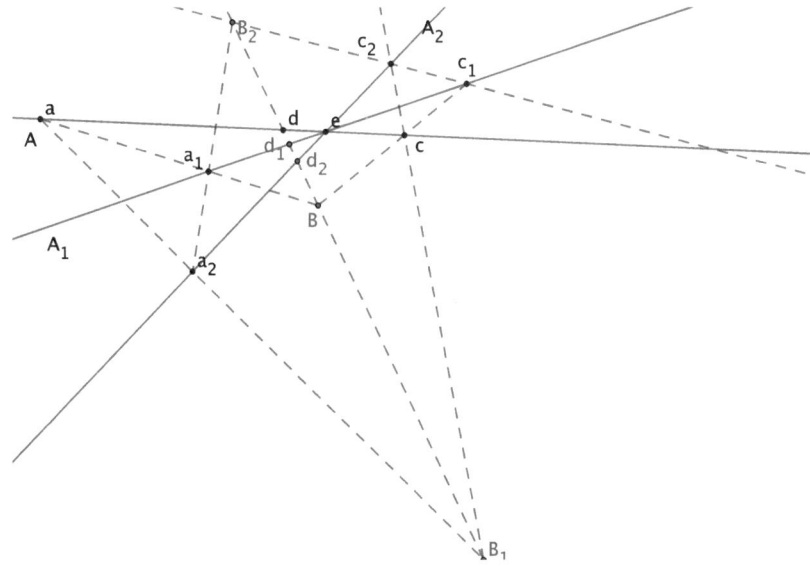

Fig. 9.2 Based on Steiner, 1832, Figure 27

Then project line A_1 to line A_2 from a point B_2: a_1 to a_2, c_1 to c_2, and e to itself. Let d be $A \cap BB_2$. So the first perspectivity maps d to point d_1 on A_1, where $d_1 = BB_2 \cap A_1$, and the second perspectivity maps d_1 to point d_2 on line A_2, where $d_2 = BB_2 \cap A_2$. So we have a projective relationship from line A to A_2 where e is paired with itself. Thus, by the La Hire-Steiner Theorem, A and A_2 are in perspective. This means that line dd_2 lies on point $aa_2 \cap cc_2$, the point we call B_1. But line dd_2 is line BB_2. Thus, B, B_1, and B_2 are collinear.

Desargues' Theorem follows: let triangles aa_1a_2 and cc_1c_2 be in perspective from point e. "Then the three points B, B_1, B_2, in which the corresponding sides, taken in the same pairwise order, meet each other on one line." [p. 293]

6. **Conic Sections** Steiner offered a new and productive approach to the conic sections, including transparent justifications for long-known constructions.

As expected by duality, there are *point conics* and *line conics*. Here we follow the point conic development. See our Fig. 9.3, Steiner's *Fig. 37*. First, recall that a conic is the image of a circle under a projectivity. Second, we take two points on the circle, B and B_1. We then pair each line on B with the line on B_1 that it meets on the given circle, for example lines a and a_1. If we take four lines on B, the cross-ratio of those lines is defined by the angles they form at B, and by the way the lines of the two pencils are paired, with the Inscribed Angle Theorem, the corresponding

Fig. 9.3 Steiner's Figure 37,
1832

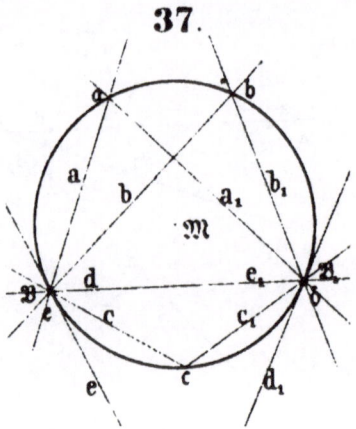

cross-ratio of the lines on $\mathcal{B}_1$ is equal. When we project the circle to a conic section, the cross-ratios are preserved, giving us the following characterization:

> **Definition** A *point conic* is the set of points of intersection of corresponding lines of two projectively related pencils.
>
> A *line conic* is the set of lines (tangents) joining corresponding points of two projectively related lines.

Steiner summarized his claim about a line conic in Art. 38 IV, in the left column, of his 1832 [98]:

Every two coplanar lines a and a_1, in an oblique [not perspective] projective relation, generate a conic section to which they are tangent, i.e., all the lines joining corresponding points are together the collection of tangents of a determined conic section . . . .

As we illustrate in our second application below, Steiner's characterization of a conic section is a clear way to justify point-by-point constructions of conics.

3   Application 1 of Projective Geometry: A Line Conic in Steiner, 1832

Steiner developed a method to produce a line conic. The argument starts with a lemma proved by Poncelet. See our Fig. 9.4, based on Steiner's *Fig*. 38 of 1832, which is nearly identical to Poncelet's *Fig*. 71 of 1822. Here is the claim:

Lemma 9.1 (Poncelet [87, Art. 464]) *Given a circle with center M and fixed tangents a and a_1, from a point P, that meet the circle at E and D_1, respectively,*

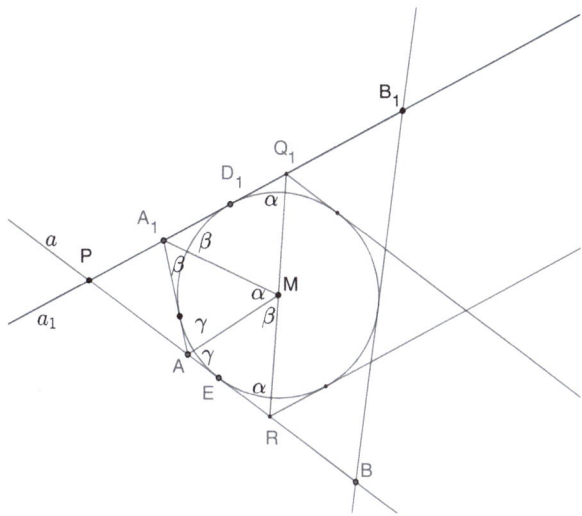

Fig. 9.4 Based on Steiner's Figure 38, 1832

draw any third tangent to the circle, as pictured, meeting tangent a_1 at A_1 and the second tangent, a, at A. Then the measure of $\angle A_1 M A$ is the same for all choices of the tangent $A_1 A$. The claim holds when we have a conic section in place of the circle and a focus of the conic in place of M.

Proof See Fig. 9.4. Mark Q_1 and R on a_1 and a, respectively, so $PQ_1 = PR$ and $Q_1 R$ is a diameter of the circle. Let α be $m\angle PRQ_1 = m\angle PQ_1 R$, and let the equal angles at A_1 have measure β and the equal angles at A have measure γ. In quadrilateral $Q_1 R A A_1$, in degrees, $2\alpha + 2\beta + 2\gamma = 360$ so $\alpha + \beta + \gamma = 180$. This means that $m\angle A_1 M A = \alpha$. Since $m\angle P = 180 - 2\alpha$, then α is constant, no matter how tangent AA_1 is selected.

Poncelet had shown with his *Fig.* 69 that there is a homology whose center is at a focus, F, of a conic section which maps the conic to a circle centered at F. (See *Application* 2 at the end of Chap. 7.) So if from an outside point P two tangents a and a_1 are drawn to a conic, and a tangent to the conic meets a in point A and meets a_1 in point A_1, then the homology maps angle $A_1 F A$ to itself. Thus, Poncelet's conclusion about a circle with the two tangents from an outside point still holds for a conic section.

Steiner showed in his own [98, *Fig.* 38], that tangent lines a and a_1 are projectively related by pairing each point A of a with point A_1 of a_1, when $A_1 A$ is tangent to the circle. In this way, B and B_1 are also paired. Steiner proved the claim by showing that cross-ratios of corresponding points are equal. (See Exercise 1 at the end of this chapter.)

We give a proof calling on properties of elementary forms in perspective. Start with the points of line a. Each point A of line a is paired with line AM, giving a projective relation of line a with the pencil on M. Then we rotate the lines on M through angle AMA_1. Rotation preserves angles, so it preserves the cross-ratio of lines on M, so it is a projective relation. Finally, a line MA_1 on M is paired with point A_1 on line a_1. Thus, lines a and a_1 are projectively related. Therefore, following Steiner's definition of a conic, the lines joining corresponding points of a and a_1 form a line conic.   □

4   Application 2 of Projective Geometry: Construction of Conics

In this section, we give justifications by concepts of projective geometry for constructions of conic sections, including well known constructions by Isaac Newton and Colin Maclaurin.

Isaac Newton (1642–1727) proved that bodies, such as the planets of the solar system, move in elliptical paths about a body, the sun, which pulls on them with a force inversely proportional to the square of their distances from the sun. The sun will be at a focus of the ellipse. In the first book of the *Principia*, Newton included much information about conic sections, including the construction of Prop. 22.

In Prop. 22 , we are given five coplanar points of a conic section: A, B, C, D, P, and asked to find more points. See Fig. 9.5. We fix two angle measures, $\angle ABC$ at B, and $\angle ACB$ at C. Draw DB and DC. Plot point M by drawing ray BM so $\angle ABC \cong \angle DBM$ and ray CM so $\angle DCM \cong \angle ACB$. Similarly, find point N so $\angle ABC \cong \angle PBN$ and $\angle PCN \cong \angle ACB$. Draw line MN. Now for any new point N on line MN, go through the same process to find the new corresponding point P on the conic section: draw ray BP so $\angle PBN \cong \angle ABC$, and draw ray CP so $\angle PCN \cong \angle ACB$, where P and N are the new points.

Fig. 9.5 Newton's
Construction of a Conic, from
Principia, 1687, Section 5
Prop. 22

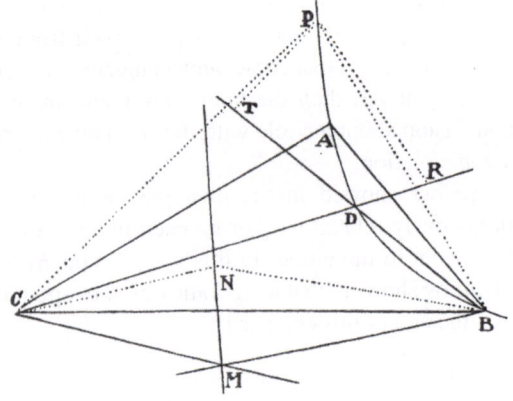

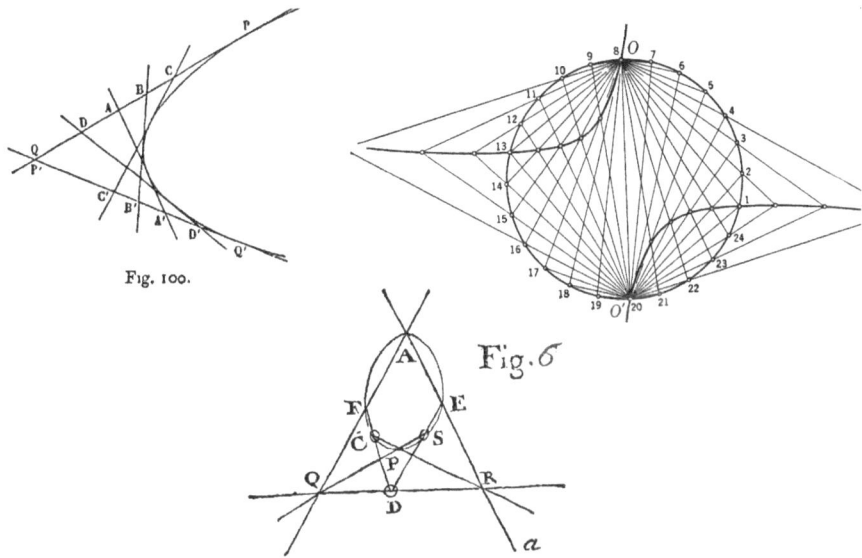

Fig. 9.6 Constructions of Conics. Upper Left: Cremona 1893. Upper Right: from *What is Mathematics?*, 1941, p. 205. Lower: Maclaurin 1735

Why do we produce a conic section? The pencils on B and C are projectively related, pairing rays BD and CD, rays BA and CA, and, generally, pairing rays BP and CP, since any pair of rays at B form an angle equal to that between the corresponding rays at C. So by Steiner's characterization, we have a point conic. (Newton offered no proof.) □

The three diagrams of Fig. 9.6 are further examples of conics formed according to Steiner's characterization. Exercise 3, at the end of the chapter, asks why the Upper Right construction produces a point conic.

The Upper Left figure is from [34] of Luigi Cremona (1830–1903). The corresponding letters are spaced proportionally, as one could get from projecting one to the other when they are drawn parallel, before they are moved into the position seen. Note that translation and rotation are projective transformations.

The lower figure is from Colin Maclaurin (1698–1746), in a 1735 article [64]. AQ and AR are given lines, and D, C and S are given points, not collinear and not on AQ or AR. We draw a line on D which meets the two given lines in Q and R, establishing a perspectivity between lines AQ and AR. Then we draw the lines SQ and CR. This creates projectively related pencils, one on C and the other on S. Where the corresponding lines CR and SQ meet, at P, we have a point of the conic. Different choices of the line on D produce different points of the conic section.

5　Application 3 of Projective Geometry: The Cross-Joins Theorem

This theorem seems inconceivable before Steiner.

Theorem 9.2 (Cross-Joins Theorem, *Steiner, 1832*) *Let two projectively related lines meet at E, with points $A, B, C, \ldots$ on one projectively related to $A_1, B_1, C_1,$ $\ldots$, respectively, of the other line. Then the cross-joins are collinear, namely, $AB_1 \cap A_1B, AC_1 \cap A_1C, BC_1 \cap B_1C, \ldots$ (Fig. 9.7).*

Proof Suppose E on line AB is not projectively related to itself.

Let E on line AB be related to E_1 on line A_1B_1. Consider the pencil with center $B_1 = \mathcal{B}_1$ of lines $B_1A, B_1B, B_1C, B_1E, \ldots$, and the pencil with center $B = \mathcal{B}$ of lines $BA_1, BB_1, BC_1, BE_1, \ldots$, with lines paired in this order. Since points $A, B, C, \ldots$ are projectively related to $A_1, B_1, C_1, \ldots$, then these pencils are projectively related in the same order. Line BB_1 is paired with itself, so by the La Hire-Steiner Theorem the two pencils are in perspective, i.e., corresponding lines meet in collinear points. The collinear points include $AB_1 \cap A_1B, BC_1 \cap B_1C, B_1E \cap BE_1$. Point $B_1E \cap BE_1$ is E_1. By a similar argument, corresponding pencils with centers A and A_1 are projectively related, showing that $AB_1 \cap A_1B, AC_1 \cap A_1C, A_1E \cap AE_1$ are collinear. $A_1E \cap AE_1$ is E_1. Thus, $AB_1 \cap A_1B, AC_1 \cap A_1C, BC_1 \cap B_1C$ and E_1 are collinear. Continuing this way, we see that all the cross-joins are collinear. (The case when E is paired with itself is an Exercise.)　　　　　　　　□

A special case of the Cross-Joins Theorem is often called Pappus's Theorem. Note that pairing A, B, C with, respectively, A', B', C' defines a projective relation of lines AB and A_1B_1, pairing E with a point E_1 of line A_1B_1.

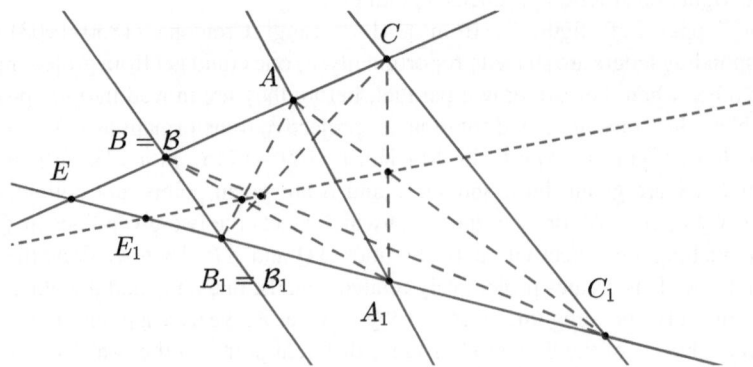

Fig. 9.7 Based on Steiner's *Fig.* 33, 1832, [98], Cross-Joins Theorem

Theorem 9.3 (Pappus's Theorem) *If the vertices of a hexagon $AB_1CA_1BC_1$ lie alternately on two lines, then the opposite sides of the hexagon meet in collinear points.*

6   Application 4 of Projective Geometry: A Problem from Poncelet 1822 Art. 495

Poncelet posed this problem: A polygon of n sides is to be constructed so the vertices lie on $n - 1$ given lines, $m_1, m_2, \ldots, m_{n-1}$, and its sides on n given points $P_1, P_2, \ldots, P_n$. Once the first vertex V_1 is selected on line m_1, all the remaining vertices are determined. Show that the locus of the final vertex V_n lies on a conic section.

Note the similarity to the problem Brianchon solved in [17], in which the points $P_1, P_2, \ldots, P_n$ are collinear. Poncelet's solution did not use projective methods. However, with Steiner's description of a conic section, projective methods provide a relatively simple solution.

Solution Consider the pencils of lines on the n points $P_1, P_2, \ldots, P_n$. Projectively relate pencil P_1 with pencil P_2 by pairing lines which meet on line m_1, relate pencils P_2 and P_3 by pairing lines which meet on line m_2, etc. In this way, pencils P_1 and P_n are projectively related. So the set of points of intersection of corresponding lines on P_1 and P_n lie on a conic. The line on P_1 and V_1 will meet its corresponding line on P_n in the final vertex V_n. □

7   The Projective Geometry of von Staudt

When Philippe de la Hire had introduced a "line [segment] divided harmonically," in 1673, he used segment length: AD is divided harmonically at B and C if $AD \cdot BC = AB \cdot CD$ and B lies between A and C. Lazare Carnot, in [22] of 1806, introduced the concept of a *complete quadrilateral*, and proved that it yielded a harmonic set. Von Staudt, rejecting concepts that depended on length, defined a harmonic set based on the complete quadrilateral, independent of segment length.

Where Steiner had defined a pairing of two forms to be *projective* when the cross-ratio is invariant, von Staudt defined a projective transformation as one which preserved harmonic sets.

Definition (Von Staudt, 1847)
See Fig. 9.8. Given lines a, b, c, d where no three are concurrent (thus forming a *complete quadrilateral*), then the line on points $X = a \cap b$ and $Y = c \cap d$

(continued)

Fig. 9.8 Points X and Y are
harmonic conjugates with
respect to U and V

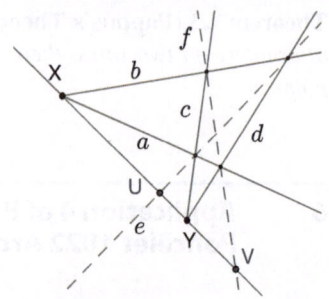

is *cut harmonically* at $U = XY \cap e$ and at $V = XY \cap f$, where e is the line on
$a \cap c$ and $b \cap d$, and f the line on $a \cap d$ and $b \cap c$. We write $H(UV, XY)$.
The dual produces a *harmonic set of concurrent lines*. Von Staudt showed that
exactly one of U and V lies between X and Y. In that sense we say U and V
separate X and Y.

[Art. 103] Two forms are *projective* when they are related so every
harmonic set of one corresponds to a harmonic set of the other.

Following on these definitions, von Staudt gave what is often called the Fundamental Theorem of Projective Geometry.

Theorem 9.4 (Fundamental Theorem of Projective Geometry) Art. 106 and
107 of [104]. *Two projectively related elementary forms are determined by the
correspondence of three elements. In particular, if two projectively related forms on
the same line have three points corresponding to themselves, i.e., are "fixed points,"
then all points of the line are fixed.*

In Steiner's geometry, the theorem was an immediate consequence of Steiner's
choice to characterize two elementary forms as projectively related when the cross-
ratios of corresponding points are equal. Why? For any three collinear points
A, B, C, and a particular value k of the cross-ratio, except 0, 1, or ∞, there is
a unique D so $CR(AB, CD) = k$. Von Staudt chose to say that two forms are
projective when harmonic sets correspond. However, his proof of the Fundamental
Theorem of Projective Geometry depended on an unacknowledged property of the
real numbers.

Here is von Staudt's proof. We suppose the elementary form is a line projectively
related to itself and that at least three points are fixed, i.e., related to themselves.
First, suppose there is a segment AB where A and B are fixed, but no point between
them is fixed. This is impossible because we are assuming some point Q outside
the segment AB is fixed, and the harmonic conjugate of Q, a fixed point, would
lie between A and B. Thus, between any two fixed points is another fixed point.

Therefore, the set of fixed points is dense throughout the line and, therefore, all points are fixed.

For the proof, von Staudt suggested that continuity is needed in that we need "*stetige Aufeinanderfolge von Elementen*" (continuous succession of elements). In 1847, the nature of the real line was not well understood. Von Staudt did not give an axiom of continuity nor further discussion of the concept. In Veblen and Young [102, p. 95] of 1910, for example, the assumption based on continuity is explicit:

Assumption P of Veblen and Young 1910 If a projectivity leaves each of three distinct points of a line invariant, it leaves every point of the line invariant.

The Fundamental Theorem, given above, follows from Assumption P. In what follows we will assume Veblen and Young's Assumption P and with it the Fundamental Theorem of Projective Geometry.

8   Foundational Issues for von Staudt

Von Staudt had postulates, just as did Euclid, but with a modern level of abstraction. In Article 18, for example, von Staudt wrote that if two points lie in a plane, then so does the line on those two points. No one before von Staudt felt such a claim needed to be made.

The type of duality emphasized by von Staudt was *point-plane duality*, where "point" and "plane" are interchanged in a statement. In the triple consisting of point-line-plane, points and planes are called *reciprocals*. In 1825 [48], Gergonne had written of both point-line duality and of point-plane duality, but after von Staudt, the latter was common in projective geometry.

Here are reciprocal statements in Art. 66, in dual pairs, to be regarded as postulates:

β_1. A line and a point not on the line determine a plane.
β_2. A line and a plane not containing the line determine a point.

In Art. 67:

γ_1. To draw a plane on three points not collinear.
γ_2. To draw a point on three planes which do not lie on one line.

Another pair:

δ_1. Two lines [in space] with a common point determine a plane.
δ_2. Two planes that meet determine a line.

These postulates sound obvious, but it is enlightening to see how von Staudt derived Desargues' Theorem [Art. 87] from them.

Theorem 9.5 (Desargues' Theorem) *Suppose we are given two triangles in space,* ABC *and* $A'B'C'$, *on distinct planes, and that their corresponding sides,* AB *and* $A'B'$, AC *and* $A'C'$, *and* BC *and* $B'C'$, *meet in collinear points on a line* u. *Then, the three lines on corresponding vertices,* AA', BB', *and* CC', *are concurrent.*

Proof By δ_2, u must be the line at which the two planes meet. By δ_1, there is a plane π_1 on A, B, B', A', a plane π_2 on A, C, C', A', and a plane π_3 on B, C, C', B'. By law γ_2, there is a point S common to the three planes, and that point must lie on the three lines of intersection of pairs of the planes. So lines AA', BB', and CC', are concurrent at S. □

As a mark of the abstract level at which von Staudt wished to work, he included no figures in his 1847 book. The story was not to depend on pictures.

---

9   Some Important Theorems in von Staudt, 1847

In Art. 128, von Staudt proved a proposition important in understanding homology.

Theorem 9.6 *If a collineation has a line of fixed points, then it has a center* S, *a point so that every line on* S *is mapped to itself. (The dual holds.)*

Proof Von Staudt made an argument involving two planes in space, but we can just use the two-dimensional Desargues' Theorem. See Fig. 9.9. Suppose we are given a collineation mapping a plane to itself, with an axis, a line of fixed points. Let XYZ and $X'Y'Z'$ be two corresponding triangles whose three pairs of corresponding sides, XY and $X'Y'$, etc. meet in points on the axis, the line of fixed points. Then, by Desargues' Theorem, the three lines on the corresponding vertices meet at a single point, S. The line XX' meets the line of fixed points at a point A; since A is fixed and X is mapped to X', then line XX', which lies on S, is mapped to itself. Likewise, line YY' is mapped to itself. So the point at which XX' and YY' meet, point S, is fixed. Any line on S meets the line of fixed points at some point B. Since S and B are fixed, the line SB is mapped to itself. So S is a center, and the given collineation is a homology. □

Fig. 9.9 Von Staudt, 1847, Figure for Art. 128. LMN is the axis (No figures are in von Staudt's work)

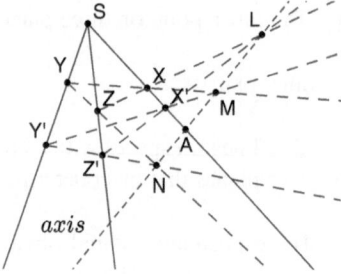

It is most interesting to see how von Staudt, in his Article 123, proved Moebius's proposition about plane-to-plane collineation. (Note that a collineation maps a complete quadrilateral to a complete quadrilateral, thus harmonic sets are mapped to harmonic sets, so a collineation is projective.) He gave two proofs.

Theorem 9.7 *A collineation of the projective plane is determined by pairing any four points, no three collinear, with any four points, no three collinear.*

Proof First, suppose a plane is to be paired pointwise with itself by a collineation, with points A, B, C, D, no three collinear, paired with, respectively, A_1, B_1, C_1, D_1, no three collinear. Line AB corresponds with line A_1B_1 and line CD corresponds with line C_1D_1 so point $E = AB \cap CD$ must be paired with point $E_1 = A_1B_1 \cap C_1D_1$. Since pairing three points of line ABE to three points of line $A_1B_1E_1$, defines a projective correspondence, then the correspondence is defined for all points of line AB. Likewise, a pairing for point $F = AC \cap BD$ is determined, so it is determined for all points of line AC. That is enough to determine the pairing for all points of the plane.

For a second proof, these lines must be paired: AB, AC, AD with, respectively, A_1B_1, A_1C_1, A_1D_1. So a projective relationship of the pencils on A and A_1 is determined. Likewise for pencils on B and B_1. So any other point X will be, generally, of form $XA \cap XB$, and, therefore, X is paired with the intersection of the lines corresponding to XA and XB. (Points on line AB can then be handled.) [Art. 130] □

Because von Staudt did not use the cross-ratio, it had to be proved that any three points in a given order determine the fourth point of a harmonic set. He stated a theorem, with proof, in his Art. 93. Our proof is from [79].

Theorem 9.8 (Three Points of a Harmonic Set Determine the Fourth Point) *Given three collinear points R, S, T, then there is a unique point U so that $H(RS, TU)$.*

Proof As in Fig. 9.10, the opposite sides of quadrilateral $ABCD$ meet in points R and S, and the diagonals AC and BD meet line RS at T and U, respectively. $A_1B_1C_1D_1$ is any other quadrilateral whose sides, corresponding to those of $ABCD$, meet RS at R, T, and S. We need to show that B_1D_1 lies on U, meaning that for given R, S, T, point U collinear with R, S, and T, is uniquely determined so $H(RS, TU)$.

Consider triangles ABC and $A_1B_1C_1$. Since corresponding sides meet in collinear points, then by Desargues' Theorem there is a point V (not shown in Fig. 9.10) at which AA_1, BB_1, and CC_1 meet. We have a corresponding result for triangles ACD and $A_1C_1D_1$. This means that AA_1, BB_1, CC_1, and DD_1 all meet at V. Since triangles ABD and $A_1B_1D_1$ are in perspective from V, their corresponding sides meet in collinear points U, T and S. So B_1D_1 lies on U. □

Fig. 9.10 Von Staudt, 1847,
Figure for Art. 93: Three
points determine the fourth in
a harmonic set

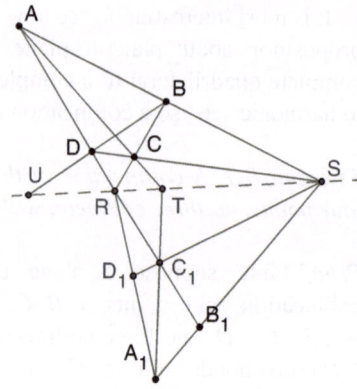

10   Application 5 of Projective Geometry: Involution, Notation, and von Staudt

Von Staudt introduced notation, important today with only minor variation, to indicate how two sets of collinear points, or two sets of concurrent lies, are projectively related. If, say, collinear points A, B, C, D are projectively related, in order, to points E, F, G, H of another or the same line, von Staudt wrote $ABCD \barwedge EFGH$.

With this notation, he could express and prove an important lemma and a theorem about *involution*.

Lemma 9.2 (1847, Art. 119) *For any four collinear points, $ABCD \barwedge BADC$.*

Proof See Fig. 9.11. Project from a center M onto a line on D, mapping A, B, C, D to, respectively, E, F, G, D. Let MC meet AF in N. Project line ED to line MC from center A. These two projections gives $ABCD \barwedge EFGD \barwedge MNGC$. Now project from center F onto the original line. We get $MNGC \barwedge BADC$. So $ABCD \barwedge BADC$. $\square$

Fig. 9.11 Von Staudt, 1847,
Figure for Art. 119

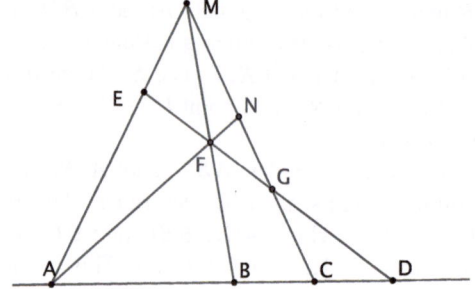

Theorem 9.9 (Art. 215) *Suppose that points $A, B, C, \ldots$ are projectively related to points of the same line, where A is mapped to A', B to B', etc. Suppose one pair is switched, say A is mapped to A' and A' is mapped to A. Then for all points B of the line, B is mapped to B' exactly when B' is mapped to B.*

Proof A projective relationship on lines or pencils is completely determined by the pairing of three points. Suppose that in this projective pairing, A and A' are mapped to each other, while B is mapped to B'. We need to show that B' is mapped to B. Let B'' denote the image of B'. So $AA'BB' \barwedge A'AB'B''$. By the Lemma, $A'AB'B'' \barwedge AA'B''B'$. Since three elements agree and the relationship is projective, $B'' = B$. Conclude that each point X is exchanged with its image X'.   □

---

11   Optional: A Taste of Projective Algebraic Geometry

Algebraic geometry is broadly concerned with polynomial functions and their graphs. For a polynomial function in projective coordinates (x, y, z) to make sense, any polynomial f that equals 0 for a triple (x, y, z) must also equal 0 for any (kx, ky, kz) when $k \neq 0$. This is only possible if the *degree* of each term $ax^l y^m z^n$, i.e., $l + m + n$, of f is the same. (A good undergraduate introduction, from which some of this material is taken, is [13].)

For this reason, if we wish to deal with a polynomial function in variables (x, y) in the projective plane, we *homogenize* the polynomial, if necessary, by adding factors of z to the terms so all terms will have the same degree.

Consider, for example, the polynomial $f(x, y) = x^3 - 2xy^2 + y^2 - 1$. The homogenized polynomial will be $F(x, y, z) = x^3 - 2xy^2 + y^2 z - z^3$. Any real root (x, y) of f corresponds to the root $(x, y, 1)$ of F, and vice versa.

A major object of study in algebraic geometry is the solution set, called a *variety*, of a system of equations. When we wish to examine the behavior of a function $f(x, y) = 0$ "at infinity," we solve the system $F(x, y, z) = 0$, $z = 0$, when F is the homogenized form of f. For our example, the system $F(x, y, z) = 0$, $z = 0$ is $x^3 - 2xy^2 = 0$. We see that the roots are $x = 0$ and $y = \pm \dfrac{x}{\sqrt{2}}$. Looking at the graph of $f(x, y) = 0$, Fig. 9.12, the lines $y = \pm \dfrac{x}{\sqrt{2}}$ appear to be asymptotes. Examining the difference of, for example, one branch of the graph, the function $\sqrt{\dfrac{x^3 - 1}{2x - 1}}$, and one asymptote, $y = \dfrac{x}{\sqrt{2}}$ for x positive and large, we verify that the difference shrinks to 0 as x goes to infinity. So we have an asymptote. As for the root $x = 0$, giving triple $(0, 1, 0)$ at infinity, this agrees with any vertical line; by the function itself we find the vertical asymptote $x = 1/2$.

As an interesting example, consider this theorem:

Fig. 9.12 Graph of
$x^3 - 2xy^2 + y^2 - 1 = 0$

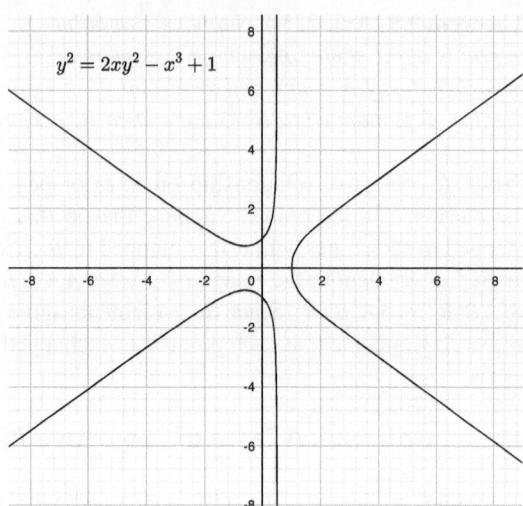

Theorem 9.10 *Let A, B, C be points of a conic; let $tan A$ denote the tangent line at A, and a, b, c the points, respectively, $tan B \cap tan C$, $tan A \cap tan C$, $tan A \cap tan B$. Then lines Aa, Bb, Cc are concurrent.*

One can prove it, for example, when the conic is a circle, by Ceva's Theorem. A conic is a projection of a circle, and a projectivity is a collineation, so the theorem holds for a conic. What happens when the conic to which the circle is projected has points at infinity?

When a circle is projected to a hyperbola, two points of the circle are projected to the points at infinity at which the asymptotes are tangents. If B and C are the two points of a circle projected to infinity, then the asymptotes are $tan B$ and $tan C$. See Fig. 9.13. We take a point A of a hyperbola, and let A' be $tan B \cap tan C$, the *center* of the hyperbola. Let B' be $tan A \cap tan C$ and C' be $tan A \cap tan B$. Take the line on C' that is parallel to $tan C$ and take the line on B' that is parallel to $tan B$, and let these two parallels meet at P. (What is the line CC'? Lines on C are parallel since

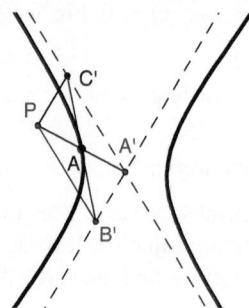

Fig. 9.13 Hyperbola, based Bix's [13, p. 81]

C is at infinity. So line CC' must be parallel to line $tan C$, and on C'.) By the above theorem, we have:

(i) Line PA lies on A',
(ii) Since $PC'A'B'$ is a parallelogram, whose diagonals bisect each other, A is the midpoint of segment $B'C'$, and A is the midpoint of segment PA'.

Apollonius had shown this in Prop. 3 of Book 2 of his *Conics* [6].

Theorem 9.11 (Apollonius) *Given a hyperbola with asymptotes c and b meeting at A'. If the tangent to the curve at a point A meets the asymptotes at B' and C', then A is the midpoint of segment $B'C'$.*

Now consider a parabola. If the equation $y = ax^2$ is homogenized, we have $yz = ax^2$. Solving with equation $z = 0$, we get a single point, a double root, $x = 0$, on the line at infinity. So the line at infinity is the tangent to the parabola on the projective plane. Now, what does the above theorem tell us about a parabola? This is an Exercise at the end of the chapter.

12 Exercises—Projective Geometry

1. From Steiner's 1832 *Systematische Entwicklung*, Art. 37. See Fig. 9.4, repeated as Fig. 9.14. Lines a and a_1 are tangents from P to the circle with center M. Q_1 and R are marked on a and a_1, so line $Q_1 R$ is a diameter of the circle and lengths PQ_1 and PR are equal. Lines AA_1 and BB_1 are tangent to circle M.
   (i) Prove that $\triangle Q_1 A_1 M \sim \triangle RMA \sim \triangle MA_1 A$.
   (ii) Prove that $AR \cdot A_1 Q_1 = (MR)^2 = (MQ_1)^2$.
   (iii) Prove that $CR(AB, R\infty) = CR(A_1 B_1, \infty Q_1)$.
   (iv) Prove that the pairing of the points of lines a and a_1 as A is paired with A_1 is projective.
2. Prove the Cross-Joins Theorem when the point E at which lines EC and EC_1 meet is paired with itself.
3. Explain why we have a point conic in the upper right diagram of Fig. 9.6, taken from *What is Mathematics?* The twenty numbered points are equally spaced around the circle.
4. (a) (From L. W. Dowling [37, p. 53], of 1917, who referenced Chasles, *Geometrie Superieure* 1880, and the *Collections* of Pappus.)

   Given two fixed straight lines u and u_1 intersecting in O, and two points S and S_1, collinear with O. A straight line rotates about a fixed point U and intersects u and u_1 in A and A_1, respectively. Show that the locus of the intersection of SA and $S_1 A_1$ is a straight line passing through O.

   (b) State the dual proposition.

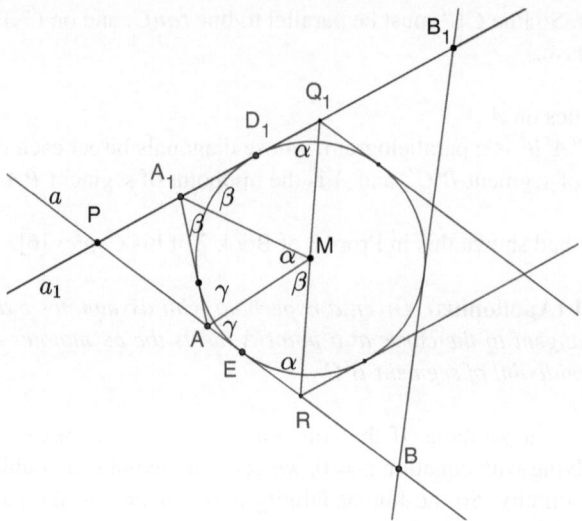

Fig. 9.14   Based on Steiner's Figure 38, 1832

5. Given four points in a plane, A, B, C, D with no three collinear. Find a point S so that the lines SA, SB, SC, SD form a harmonic set.

6. Explain: Steiner's characterization of a projective relationship by the invariance of the cross-ratio, and by his characterization of a point conic as the set of points of intersection of two projectively related pencils, lets us show that "There is a unique conic section on any five points".

    Note: This claim is true if no three of the five points are collinear. Poncelet wrote that a conic section is the projection of a circle in one plane onto another plane. As no line can meet a circle in three points, no line could meet a conic section in three points. However, if we think of a conic section as the solution set of an equation of form $Ax^2 + By^2 + Cxy + Dx + Ey + F = 0$, then there are many conic sections on four collinear points and another, fifth, point.

7. (Based on the Optional section, on projective algebraic geometry.) Let points A and B be on a parabola and C the point at infinity of the parabola. What is the claim of the theorem of that section? Prove it. (Note that $tanC$ is the line at infinity.)

8. We have learned that there is a unique collineation of the projective plane mapping any four points, no three collinear, to any four points, no three collinear. Let $ABA'B'$ be a parallelogram and α the collineation that interchanges A and A', and interchanges B and B'. Show that α is a homology by showing that it can be described by a *center* and *axis*. What is the center? What is the axis?

Transformation in German Universities

10

We saw earlier that the long period of reform and revolution in France, roughly from 1760 to 1820, was accompanied by transformation in the institutions of mathematics: the schools that trained mathematicians and the forms of communication in mathematics. That transformation was followed by a transformation centered on German universities. The German transformation was sudden and it moved the center of higher level mathematics from France to Germany. The appearance of *Crelle's Journal, Journal für die reine und angewandt Mathematik*, in 1826, can be taken as the point of the switch.

Where *Gergonne's Journal* was closely tied to the *École Polytechnique*, and emphasized geometry, *Crelle's Journal* was not tied to one institution and published articles on a broad range of mathematics, from a broad range of mathematicians. Its first issues were filled with pieces by Carl Jacobi, Niels Abel, Jacob Steiner, and August Moebius, the young pioneers in algebra, elliptic integrals, and geometry. Before, groundbreaking mathematics had generally appeared in books, such as Descartes' *Géométrie*, Newton's *Principia Mathematica*, and Gauss's *Disquisitiones Arithmeticae* of 1801, which revolutionized algebra, physics, and number theory. With *Crelle's Journal* new ideas were as likely to appear in journal articles, which remains true today.

France still produced great mathematicians, such as Augustin-Louis Cauchy (1789–1857) and Joseph Liouville (1809–1882), and French institutions of education were admired and imitated throughout the world, but the emphasis was on the applications of mathematics. Exciting ideas came more often from Germany. The story of Poncelet is illuminating. He was a military officer and, after the Napoleonic Wars, taught at military academies, even as he produced his *Traité* of 1822 and, before 1830, a number of papers on projective geometry. In 1824 he accepted the position of Professor of Mechanics at Metz, and, in 1832, Professor of Mechanics at the Sorbonne. As reported in the MacTutor history [75], in the 1820s "he applied mechanics to improve turbines and waterwheels more than doubling the efficiency of waterwheels." He would be director of the *École Polytechnique* from 1848 to

1850, at the same time leading troops in putting down the rebellion of 1848. But his days of innovation in mathematics ended long before 1848.

What happened in Germany? One writer put it this way: except for Carl Friedrich Gauss, in the first two decades of the nineteenth century "there was no man at any German university who was either a great teacher or an eminent researcher." [29, p. 168]. Surveying the people who were to become eminent researchers, nearly all attended German *gymnasia*, high level academic high schools, usually with at least one instructor who was a strong mathematician and an inspiring teacher. In the 1820s, strong departments of mathematics, astronomy and physics were built at several universities.

The university at Göttingen had been founded in 1734, intended to be an alternative to the medieval lethargy that still prevailed at the time. It taught modern law and was a center for science in the late eighteenth century. Gauss (1777–1855) was a student at Göttingen University. He returned there in 1807 and stayed until his death, in charge of the observatory and lecturing in mathematics. Both A. F. Moebius and K. G. C. von Staudt were students at Göttingen before 1820. Gauss would be followed at Göttingen by a succession of eminent mathematicians, including Bernhard Riemann, leading to David Hilbert (1862–1943), who is second only to Gauss in importance to modern mathematics.

The University of Berlin, founded in 1810, had the strongest mathematics faculty in the latter half of the nineteenth century. It succeeded a prestigious academy at Berlin. The Berlin-Brandenburg Society of Scientists began in 1700, with Gottfried Wilhelm Leibniz as its first president. Prussia's Frederick II revived the society in 1740, named in 1743 *Académie Royale des Sciences et Belles Lettres*. Note the French name. Euler was the first director of the mathematics section, followed by Lagrange. It was a research institute, the most prestigious at the time, but not a teaching institution. That academy eventually lost its luster; after Prussia's defeat by Napoleon, the University of Berlin was started in 1810. It had high-minded research goals. As in France, it would serve to meet the expanded need to train teachers and civil servants, but research came first. Before 1825, Jacob Steiner, Franz Neumann, and Carl Jacobi were students at Berlin. (Neumann and Jacobi, with Wilhelm Bessel, would go on to make the University of Königsberg an important center for mathematics, physics and astronomy.) Steiner and Lejeune Dirichlet were professors at Berlin, Steiner 1834–1863, Dirichlet 1828–1855. Karl Weierstrass, who can be said to have made analysis the subject we study today, was at Berlin 1856–1890. His colleagues included Ernst Kummer and Leopold Kronecker, and his students formed the core of the next generation of mathematicians.

In the 1930s, the rise of the Nazis put an end to German dominance in mathematics.

Geometric Inversion

11

1 Geometric Inversion

In the transformations considered so far, we have worked in either the real plane, represented as the set of ordered pairs of real numbers, or the extension of the real plane to the real projective plane, where one point at infinity is added to every finite line, with the collection of those points at infinity forming, itself, the line at infinity. Now we will study two other transformations in which there is just one point at infinity, a point shared by all lines. This is the plane that H. S. M. Coxeter called the *inversive plane* [30, p. 84], the plane of the stereographic projection as employed by Ptolemy, a plane whose points are in one-to-one correspondence with points of a sphere if projecting a sphere from the South Pole onto the tangent plane at the North Pole.

> **Definition**
> The *inversive plane* is the real plane to which is added one point *at infinity*. Unlike the real plane, the specially designated subsets are not just lines but lines and circles.
>
> The *inversion transformation* maps the plane to itself in such a way that lines and circles are mapped to lines and circles. Unlike the real plane, where there is exactly one line on any two points, in the inversive plane, on any three points there is one line or circle. The lines are those subsets which include the point at infinity. In recently developed vocabulary, lines and circles in the inversive plane are together referred to as *generalized circles* or *clines*.

We use the name *inversive plane* because it is the plane in which the *inversion transformation* naturally operates. We will later see the *Moebius transformation*, which operates in the corresponding plane of the complex numbers augmented by a single point at infinity. There, too, circles and lines are mapped to circles and lines.

Our treatment of inversion is based on work of Giusto Bellavitis (1803–1880). He introduced geometric inversion in a paper of 1836. Earlier work with inversion had been done in the 1820s, but Bellavitis was the first to treat the subject along the lines followed today in college geometry. We include Bellavitis's application to a construction problem: to construct a circle on two given points and tangent to a given circle.

The earliest paper involving inversion was from 1822, by J. B. Durrande. Durrande developed *conjugate poles*, *pôles conjugués*, with respect to a circle with center I [38]. We take a point A outside a circle, then construct the polar of A, meeting the circle at U and V. Where this polar meets the line IA, at A', we declare A and A' to be *inverses with respect to the circle*. One can directly show that the product of lengths $IA \cdot IA'$ is the square of the radius of the circle. There were several other works in the 1820s that included geometric inversion, including the property that lines and circles are mapped to lines and circles. Germinal Dandelin (1794–1847) connected stereographic projection to map making and Ptolemy's work; Adolphe Quetelet (1796–1874) gave the formula

$$\left(\frac{xr^2}{x^2 + y^2}, \frac{yr^2}{x^2 + y^2},\right)$$

for the inversion of a point (x, y) with respect to the circle $x^2 + y^2 = r^2$. Even Jacob Steiner had work on inversion in 1824, not published until years later. See [78] for this early history.

---

2   Introduction: Bellavitis and His Paper of 1836

Giusto Bellavitis was from Bassano. Bassano was a city within the Republic of Venice until the Napoleonic Wars; after 1815 it was included in the Kingdom of Lombardy-Venetia, which itself was incorporated into the unified Kingdom of Italy in 1866. Giusto's education began at home under the tutelage of his father, an accountant and member of the minor nobility of Bassano, after which Giusto learned mathematics on his own. He was a major contributor to *Annali delle scienze del regno Lombardo Veneto* from its inception in 1831. Only in 1843 did he take a university position, first at Vincenza, then at Padua.

The first articles by Bellavitis show an interest in complex numbers. He introduced an algebra of vectors, not using that name, which he characterized by length, inclination to a given line, and direction. Two vectors with a common length, inclination, and direction were called *equipollent*. He represented a vector in the manner of complex numbers, as $m\epsilon^\mu$, where m, the length, is real and μ is the angle at which the vector is inclined to a specified line. He defined multiplication and division of vectors corresponding to multiplication and division of complex numbers. Bellavitis used a special symbol to place between equipollent expressions, for which we use "$\equiv$."

Bellavitis used the vector notation $m\epsilon^\mu$ as we now use the complex number notation $me^{i\mu}$, with $e^{i\mu}$ denoting the complex number $cos\ \mu + i\ sin\ \mu$. Since $cos^2\mu + sin^2\mu = 1$, then $e^{i\mu}$ is a vector from the origin to a point P on the unit circle in the complex plane and μ is the measure of angle POX, where O is the origin and X the point 1, on the real axis. With trigonometric identities, $e^{i\mu}\cdot e^{i\nu} = (cos\ \mu + i\ sin\ \mu)(cos\ \nu + i\ sin\ \nu) = (cos(\mu+\nu) + i\ sin(\mu+\nu)) = e^{i(\mu+\nu)}$. Likewise, $\dfrac{e^{i\mu}}{e^{i\nu}} = e^{i(\mu-\nu)}$.

His article *Teoria delle figure inverse, e loro uso nella Geometria elementare* [10] appeared in the 1836 volume of the *Annali*. Along with further development of the algebra of vectors, Bellavitis defined the inversion transformation in the plane, developed its properties, and gave applications to construction problems. It was the first thorough treatment of inversion geometry and would not be matched in its clear simplicity for many decades.

Articles 3 and 4 provide a definition and important formulas.

Definition
Given points $A, B, \ldots$ and the circle of center I and radius r, we call I the *center of inversion* if on rays $IA, IB, \ldots$ are taken points $A', B', \ldots$, respectively, where, in length,

$$IA' = \frac{r^2}{IA}, \quad IB' = \frac{r^2}{IB}, \ldots.$$

The points $A', B', \ldots$ are called the *inverses* of $A, B. \ldots$, and vice versa, and the entire figures $A'B'C'. \ldots$ and $ABC. \ldots$ are the *inverses* of each other. r is the *radius of inversion*.

Formulas With I still indicating the center, when A and B are two other points, first,

$$AB \equiv IB - IA = \frac{r^2}{IB'} - \frac{r^2}{IA'} = r^2\frac{IA' - IB'}{IA' \cdot IB'} = r^2\frac{B'A'}{IA' \cdot IB'}. \tag{11.1}$$

The equal signs refer only to length. Further, again involving length, following from (11.1), we have

$$\frac{BM}{AM} = \frac{B'M' \cdot A'I}{A'M' \cdot B'I}. \tag{11.2}$$

3   Inversion of Lines and Circles

Articles 7 through 12 deal with inversion of lines and circles. A line on the center
is, of course, paired with itself.

How about a line EN that does not lie on the center? See Fig. 11.1 Left. I is the
center of the inversion. We will treat E as fixed, the point on the line that is closest
to the center, I, while N runs along the line. Let H be the point on ray IE where
E is the midpoint of segment HI. This means that line EN is the set of points
equidistant from I and H. By Art. 4, Eq. (11.1)

$$HN = r^2 \frac{N'H'}{IN' \cdot IH'}.$$

Now, $\dfrac{r^2}{IN'} = IN$. And $IN = HN$. So we have the equation of lengths $HN =$
$\dfrac{N'H' \cdot HN}{IH'}$. When we cancel HN and note that IH' is a constant, not depending
on N, we arrive at the equation $H'N' = IH'$. Remember that N' lies on ray IN.
This means that N' runs along the circle with center H' and radius IH'. It follows
that I is on the image circle. As N moves farther and farther from E along line
EN, the inverse point N' approaches I. As the inversion transformation is its own
inverse, our argument above means that a circle on the center, I, has a line as its
inverse. In summary:

Theorem 11.1 *For an inversion with center I, a line n not on I is paired with the
circle on I tangent, on the side of n, to the line on I that is parallel to n. If one or
two points on n are on the circle of inversion, those points are fixed. A line on I is
paired with itself.*

In Art. 10, Bellavitis asks for the inversion of a circle that is not on the center of
inversion, I. Let AB be the diameter of the given circle which lies on I. Let D and
K, on line AB, be any pair of harmonic conjugate points with respect to the given
circle. We can see by Fig. 11.1 Right that $H(KD, AB)$, i.e., K and D are harmonic
conjugates of A and B. This means that $DA : KA = DB : KB$. Now, a Circle
of Apollonius is a circle defined as the set of points X satisfying $KX : DX = e$
where K and D are given points and e is a given positive ratio not equal to 1.
(The Circle of Apollonius is treated in Exercise 1 of this chapter. It was not used
in the argument of Bellavitis.) The set of points M such that $DM : KM$ equals
the constant $DA : KA = DB : KB$ is the given circle since A and B satisfy the
equation and the set of points M is symmetric with respect to line DK. Under the
inversion with center I and radius of inversion r, by (11.2),

$$\frac{DM}{KM} = \frac{D'M' \cdot K'I}{K'M' \cdot D'I}.$$

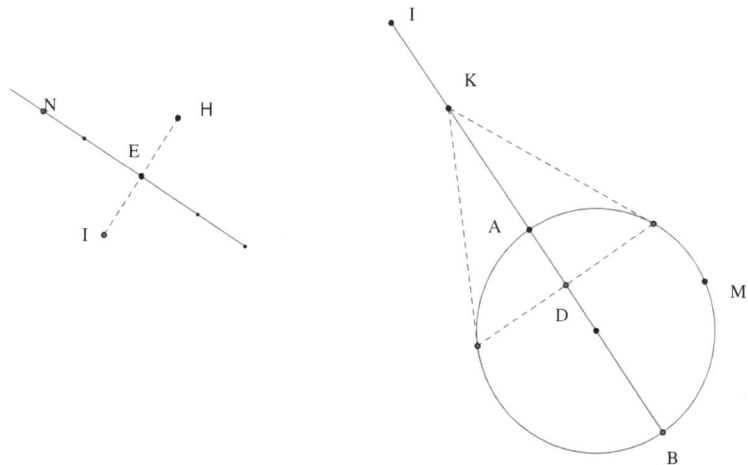

Fig. 11.1 Left: Line NE is set of points equidistant from I and H. Right: K and D are harmonic conjugates with respect to A and B; the circle with diameter AB is the set of points M satisfying $KM : DM = constant$

Now, $\dfrac{DM}{KM}$ and $K'I$ and $D'I$ are constants. Thus, $\dfrac{D'M'}{K'M'}$ is a constant. So M' lies on a circle, the circle with diameter $A'B'$.

We summarize:

Theorem 11.2 *Inversion pairs a circle not on I with a circle not on I. Let the given circle have diameter (extended if necessary) AB that lies on I. (A and B are on the given circle.) The given circle is paired with the circle whose diameter is $A'B'$.*

Bellavitis made an argument in Articles 5, 6, and 18 that the angle at which two vectors meet is invariant under inversion. Such a mapping is called *conformal*. We will give a simpler argument, found in H. S. M. Coxeter's *Introduction to Geometry* [30, p. 84], starting with a lemma.

Lemma 11.1 *Under an inversion with center I mapping a point A to A', any circle on A and A' is mapped to itself.*

Proof The circle of inversion must pass between A and A' so it meets the given circle in two fixed points X and Y. The given circle is paired with the circle on A, X, A' and Y, which is the same circle.   □

Theorem 11.3 *Let two lines or curves meet at point A, where the lines themselves or tangents to the curves are m_1 and m_2. Then the lines or circles m'_1 and m'_2 meet at A' in the same angle as m_1 and m_2 meet at A, although in the opposite sense.*

Fig. 11.2 Curves meet at A and A' in equal but opposite angles

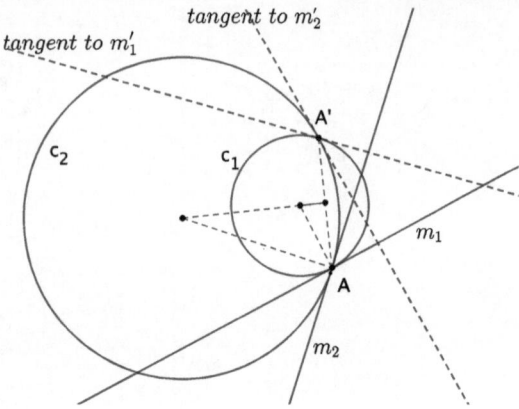

Proof See Fig. 11.2. Take the circle c_1 which lies on points A and A' and is tangent to line m_1 at A. In the same way, create circle c_2 on A and A' and tangent to m_2 at A. By the symmetry of circles c_1 and c_2 about the perpendicular bisector of AA', tangents to circles c_1 and c_2 meet at the same angle at A and at A', although in the opposite sense. (If a line m is tangent to a circle at a point A, then the image m' is tangent to the image of the circle at A'.) In this sense, angles are preserved.   □

4   Application 1 of Inversion: A Construction Problem from Bellavitis, 1836

In a modern treatment of inversion as found in college geometry texts or *What is Mathematics?*, an important application is to straightedge/compass construction problems. Bellavitis seems to have been the first to make such an application. Here is his Art. 20. See Fig. 11.3. We have seen this construction problem before as **Construction C**.

> We'll see the application of inversion to some problems. Construct a circle which passes through two given points A B and is tangent to a given circle Db. We take one of the given points, A, as the center of an inversion. Forming the inverse figure of the proposed problem will reduce it to a much easier one: to construct a line which passes through a given point B' and is tangent to the given circle $D'b'$. To make it easier we can suppose that the radius of the inversion be such that the given circle is its own inverse. Thus by means of this circle it will be easy to determine the inverse point B' of B. After the tangent $B'D'$ is drawn, it remains to construct the inverse circle of this line $B'D'$, which can be carried out by drawing AD' until it meets, again, the given circle at D, then describing the circle which passes through points A B and is tangent to the given circle at D.

Note that in Fig. 11.3, we draw the tangent from center A of the inversion, meeting the given circle at $b = b'$. The circle with center A and on b will be the circle of inversion. B' is found and a tangent from B' to the given circle is drawn, meeting the circle at D'. D is the other point of the given circle lying on line AD'.

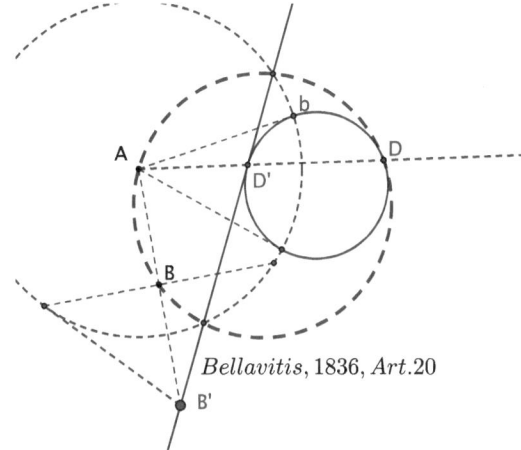

Fig. 11.3 Bellavitis Art. 20, 1836: Construct a circle on given A, B, and tangent to circle Db

By the lemma of the last section, the given circle is mapped to itself by the inversion. The line $B'D'$ is mapped to the circle on A, B, and D, which must be tangent to the given circle at D.

A systematic listing of constructions, several involving inversion, appears in Application 3.

5 Application 2 of Inversion: Steiner's Porism

In his *Geometrische Lehrsätze*, of 1827 [97], Jacob Steiner stated and proved what has come to be called *Steiner's Porism*. (*Porism* is a term going back to Pappus's work, applying the term to a lost work of Euclid. We could say that it is a claim for the existence of a figure with certain properties at the same time showing how to construct it.) Very freely translating from Article 11, referring to Steiner's Fig. 11.6, we have

Theorem 11.4 (Steiner's Porism)

> *For given circles K and K_1 that do not meet, when there is a sequence of circles $m, m_1, m_2, \ldots$ [or spheres, as Steiner framed it] so each is tangent to circles K and K_1 and to the next one in the sequence, in a way that is* commensurable, *i.e., completes the path between the two given circles, then if moved continuously between the given circles the sequence will return the first circle, m, to its original position.*

In other words, for given circles K and K_1 there is not always such a sequence of circles $m, m_1, m_2, \ldots$ filling the space between circles K and K_1. But when there is such a sequence, then starting with any circle m tangent to K and K_1, we can product such a *commensurable* sequence (Fig. 11.4).

Fig. 11.4 Steiner's *Fig.* 6, 1827

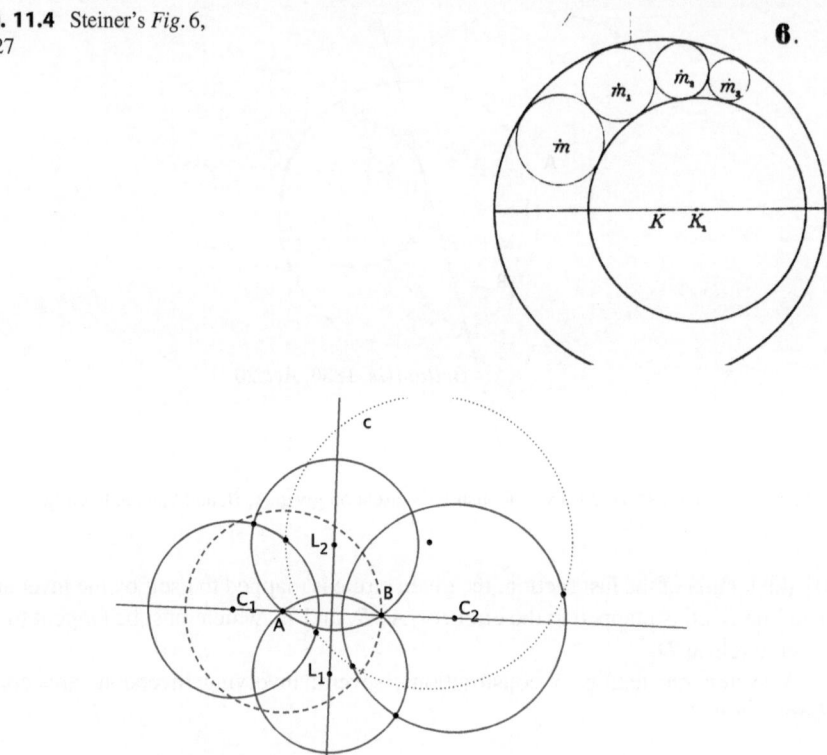

Fig. 11.5 Forder's proof of Steiner's Porism

Steiner's work does not use inversion, but inversion does provide an effective method of proving Steiner's Porism. Here we follow the method of H. G. Forder [45].

Proof Let C_1 and C_2 be circles. See Fig. 11.5, where it is easier to follow the argument when neither circle is inside the other. Our plan is to find an inversion that maps circles C_1 and C_2 to concentric circles. In the case of concentric circles, it is easy to see whether there is a commensurable sequence of circles $m, m_1, m_2, \ldots$ filling the space between the images of circles C_1 and C_2. And if there is such a commensurable sequence, the first circle of the sequence can be placed anywhere between those image circles.

We first find the common secant by drawing some circle, c, meeting both circles. (This is an application of Steiner's Common Secant Construction [96, Article 4].) We draw the common secant of circles c and C_1, and the common secant of c and C_2; the point at which those common secants meet, L_1, is on the common secant of C_1 and C_2. Let L_2 be another point on the common secant. From L_1 and L_2 we draw tangents to the circles, and then circles with centers at, respectively, L_1 and

L_2, and on the points of tangency. This gives two circles orthogonal to circles C_1 and C_2. ("Orthogonal to" means "perpendicular to.") All circles with centers on the common secant and orthogonal to circles C_1 and C_2 lie on two certain points, points A and B in Fig. 11.5. To see this, note that line C_1C_2 is the common secant for circles L_1 and L_2 (for tangents from C_1 to circles L_1 and L_2 are equal). Define A and B as the points at which circle L_1 meets line C_1C_2. From A and B, any tangent to circle L_1 has length 0. So from A any tangent to a circle such as L_2 also must have length 0. So all possible circles L_2 lie on A (and on B).

Now consider the inversion with center of inversion A where the circle of inversion lies on B. B is fixed under such an inversion. All circles and lines on A are mapped to lines, since A is mapped to infinity. So circles L_1 and L_2, and line AB are all mapped to lines on B. Further, any circle, such as C_1 or C_2, not on A and orthogonal to circles or lines on A will be mapped to a circle orthogonal to the images of circles L_1 and L_2. Any circle orthogonal to two different lines on B must have B as its center. We conclude that circles C_1 and C_2 are mapped to circles C_1' and C_2' concentric about B. We can inscribe circles tangent to C_1' and C_2'. Such tangent circles will completely fill the space between C_1' and C_2' exactly when the angle formed at B by tangents to one of the circles divides 360 (degrees). ☐

6 Application 3. Ten Related Construction Problems

Here is a construction problem amenable to solution by inversion.

Construction N Given a line, LM, a circle not on the line, and a point, A, outside the circle and not on the line, construct a circle on the point A and tangent to both the line and the circle.

Suggestion Draw tangents to the circle from A, and then draw the circle with center A and on the points of tangency. Inversion in this circle makes it a problem we have already handled.

This construction problem is found in (Poncelet 1813, *Cahier* 1, Problem 6). (Poncelet did not solve the problem by inversion.) We refer to it as the Point-Line-Circle problem.

This problem is one of a well known group of ten problems in which three geometric figures are given, selected from Point, Line, and Circle, and one must construct a circle tangent to the three figures. ("Tangent" to a point means lying on the point.) The suggested solution of this problem employs the two most common strategies of transformation by inversion: a point is mapped to ∞, and a line is mapped to a circle or vice-versa.

Here is a review of the ten problems.

Point-Point-Point is *Constructions I* in the section Exercises - Dilations.

Bellavitis's Article 20 is the *Point-Point-Circle* problem, our *Construction C*, which can be transformed by inversion to a *Point-Point-Line* problem, *Construction C₁*. Both these constructions are in the section Exercises—Greek Background.

The problem just given above is the *Point-Line-Circle* problem.

The Problem of Apollonius is the *Circle-Circle-Circle* problem, *Construction F*, and we see from Poncelet's solution that it can includes a solution of the *Point-Circle-Circle* problem, which could, itself, be transformed to a *Point-Line-Circle* problem by inversion.

Construction J is the *Line-Line-Circle* problem, and *Construction H* is the Point-Line-Line problem, both treated in the section Exercises—Dilations.

The *Line-Line-Line* problem is the well known problem of inscribing a circle in a triangle, which can be found in many high school geometry texts. (The three angle bisectors meet at the center of the desired circle.)

The *Line-Circle-Circle* problem can be handled with the same preliminary step used by Viète in solving the problem of Apollonius, and used by Poncelet in his 1809 solution of the same problem: reduce the radius of the circles by the radius, r, of the smaller circle, and translate the line by the same amount. This produces a *Point-Line-Circle* figure, which we can solve as Construction N. Then the radius of the circle found is reduced by r.

---

7   Exercises—Inversion

1. Circle of Apollonius. Given points A and B and a positive real number $e \neq 1$, show that the set of points X such that $\dfrac{XA}{XB} = e$ is a circle. Suggestion: Let A and B be $(a, 0)$ and $(b, 0)$, respectively, and $X = (x, y)$. Show the equation of the definition reduces to the equation of a circle.

2. Let A be a point outside a circle, where tangents from A meet the circle at B and C. What is the image of this figure under the inversion with center A that maps point B to itself?

3. Given a line tangent to a circle at point X. What is the image of the figure under an inversion with center X?

4. Let A and A' be inverse points with respect to a circle with radius r and center of inversion I. Prove that any circle on A and A' meets the circle of inversion at a right angle.

5. From Hadamard p. 212 Art. 215 [52]. Show that two figures found by two different inversions to the same figure, where the center of inversion is the same, are related by a dilation. Find the scale factor of the dilation.

Fig. 11.6 Inversion
Exercise 7

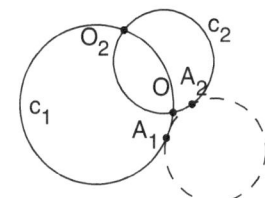

Fig. 11.7 Circles c_3 and c_4
related by the dilation with
center C pairing points B
and D

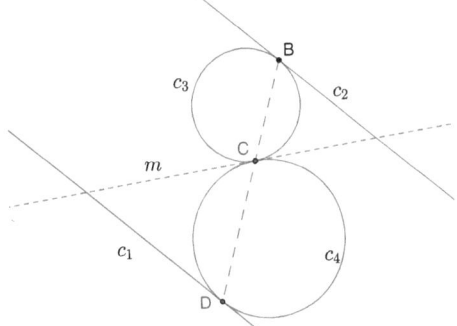

6. Hadamard p. 217 Art. 226. Given a circle and a line not tangent to the circle, find all the inversions that map the circle to the given line.

7. We are given two circles of different sizes, c_1 and c_2, which meet in two points, O and O_2. Let A_1 be a point of c_1 which is outside circle c_2. Find point A_2 of circle c_2 so there is a third circle tangent to c_1 at point A_1 and tangent to circle c_2 at A_2. See Fig. 11.6.
   Solve by inversion.

8. Prove this lemma, which may help in the solution to the problem that follows.

Lemma 11.2 *Let circles c_3 and c_4 be tangent to each other at C, and tangent to line m at point C. Let a line on C meet c_3 at B and meet c_4 at D. Show that the dilation ϕ with center C that maps B to D*

 (i) *maps circle c_3 to circle c_4,*
 (ii) *maps m to itself, and*
 (iii) *if a line on C meets circle c_3 at B and meets circle c_4 at D, then the tangent lines at B and D are parallel (Fig. 11.7).*

9. (From I. M. Yaglom, *Geometric Transformations* IV [107]). Given a sequence of four circles, the first two tangent at A, the second and third tangent at B, the third and fourth tangent at C, and the first and last tangent at D. Prove $ABCD$

Fig. 11.8 Cyclic
quadrilateral problem from
Yaglom, *Geometric
Transformations* IV, p. 15

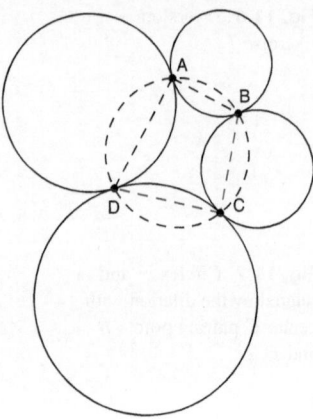

is a cyclic quadrilateral (Fig. 11.8). (Yaglom's solution is to apply an inversion
with center *A*.)

10. Solve the Problem of Apollonius by an inversion. A helpful strategy is to
    increase the radii of the given circles until two circles become tangent to each
    other, then invert.

Moebius Transformation

12

1   Moebius Transformation

As now presented in books on complex analysis or conformal mapping, a *Moebius transformation* is a function of a complex variable, whose value may be infinity, of the form

$$f(z) = \frac{az + b}{cz + d}, \quad \text{where } ad - bc \neq 0. \tag{12.1}$$

The requirement that $ad - bc \neq 0$ guarantees the existence of the inverse transform.

The first development of this transformation, by A. F. Moebius, looked very different. Modern treatments, found in books on complex number analysis or in more specialized areas with titles such as "conformal representations," first prove two important properties: that the Moebius transformation is *conformal*, i.e., maps angles to congruent angles, and that lines and circles are mapped to lines and circles. As an example or exercise, one shows that the cross-ratio is preserved by the Moebius transformation. Moebius proved those same claims, but preservation of the cross-ratio came first.

We list important definitions, some of which we have already seen.

> **Definitions**
> A *complex number* is one that can be written in form $X + iY$ where X and Y are real and i is a number whose square is -1.
>   The *complex plane*, denoted $\mathbb{C}$, is the set of all complex numbers, where the lines are the solution sets of equations of the form $ax + by + c = 0$, for

(continued)

© The Author(s), under exclusive license to Springer Nature Switzerland AG 2025
C. Baltus, *Geometry by Its Transformations*, Compact Textbooks in Mathematics,
https://doi.org/10.1007/978-3-031-72281-3_12

complex constants a, b, c where a and b are not both zero. There is one *point at infinity*, which lies on all lines. Like the plane of the inversion transform, this is a type of *inversive plane*.

The *cross-ratio* of four complex points A, B, C, D, where order matters, is

$$\frac{AC \cdot BD}{AD \cdot BC}, \text{ denoted } CR(AB, CD).$$

In the fraction, AC denotes the difference $C - A$.

Moebius's initial version, of 1853, of the Moebius transformation, which he called the *new relationship, neue Verwandtschaft*, fixed three distinct points, A, B, C, and their corresponding points A', B', C', and simply defined D' to correspond to a given point D by

$$\frac{AC \cdot BD}{AD \cdot BC} = \frac{A'C' \cdot B'D'}{A'D' \cdot B'C'}, \text{ i.e., } CR(AB, CD) = CR(A'B', C'D').$$

As Moebius wrote, "The equality of the cross-ratio (*Doppelverhältniss*) is the essence of the transformation." [68, p. 209].

One can check that this formula finds D' as the solution of a linear equation.

Moebius's definition of the function can be put into the form (12.1). Why? Starting with the equality of the cross-ratio, which defines output D' for input D, $\frac{AD}{BD} = k \cdot \frac{A'D'}{B'D'}$, where the constant k is not 0 and involves A, B, C, A', B', C' but not D or D'. Replacing AD by $D - A$, etc., we can solve for D' in form $\frac{\alpha D + \beta}{\gamma D + \delta}$.

In the case of four collinear points, claims for invariance of the cross-ratio under a specialized transformation, a line-to-line projection, go back to Brianchon in 1817 [19, p. 7]. We presented a proof in Chap. 5. Moebius gave a proof of this invariance in 1827, Art. 188. Moebius in 1853 was the first to argue for invariance involving any four points in the complex plane.

Let us present a modern argument for invariance of the cross-ratio.

Theorem 12.1 *Let w, x, y, z be four complex numbers in the plane, which need not be collinear, mapped to w', x', y', z' respectively, by a Moebius transformation of form (12.1). Then the cross-ratio is invariant, i.e.,*

$$\frac{wy \cdot xz}{wz \cdot xy} = \frac{w'y' \cdot x'z'}{w'z' \cdot x'y'} \text{ where } wy \text{ denotes } y - w, \text{ etc.} \tag{12.2}$$

Proof This argument is from Constantin Carathéodorie, 1932 [20, p. 6]. For any four complex numbers a, b, c, d where $ad - bc \neq 0$, for complex numbers w and y,

$$y' - w' = \frac{ay + b}{cy + d} - \frac{aw + b}{cw + d} = \frac{ayd - awd + bcw - bcy}{(cy + d)(cw + d)}$$

$$= \frac{ad - bc}{(cy + d)(cw + d)} \cdot (y - w).$$

When one makes a similar computation four times and then simplifies, we see that equation (12.2) holds. □

2   Complex Numbers and Geometry

In his exposition, Moebius observed that any complex number z can be represented in form

$$z = a\phi(\alpha) \quad \text{where} \quad \phi(\alpha) = cos(\alpha) + \sqrt{-1}\, sin(\alpha),$$

and a is real, usually positive, generally requiring that α be positive. In this equation, α is the modern $arg(z)$; we note that any particular α can be replaced by another value differing from α by an integer multiple of 2π. In the complex plane, $\phi(\alpha)$ represents both a point, P, on the unit circle and the vector OP where O is the origin. Vector OP meets the positive real axis at angle α. See Fig. 12.1. $(cos(\alpha) + \sqrt{-1}\, sin(\alpha)$ is now represented as $e^{i\alpha}$.) The use of complex numbers in this form was not new with Moebius; predecessors included Leonhard Euler [40].

About **Angle Measure**: The measure of angle AMB is the signed measure of the rotation about M that brings vector MA to rest on vector MB. Therefore if MA is the vector $cos(\alpha) + \sqrt{-1}\, sin(\alpha)$ and MB is the vector $cos(\beta) + \sqrt{-1}\, sin(\beta)$,

$$\angle AMB \text{ has measure } \beta - \alpha = arg\frac{MB}{MA}.$$

Fig. 12.1 Left: Complex unit vector $OP = cos(\alpha) + \sqrt{-1}sin(\alpha)$. Right: Circle above chord AB is $\{X : m\angle AXB = \beta\}$

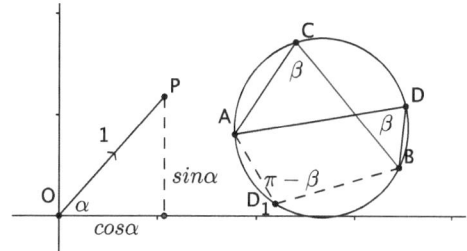

Now consider the argument of a cross-ratio $CR(AB, CD)$. It is

$$arg\frac{AC}{AD} \cdot \frac{BD}{BC} = arg\frac{AC}{BC} - arg\frac{AD}{BD} = m\angle BCA - m\angle BDA. \qquad (12.3)$$

This gives us the following lemma and theorem.

Lemma 12.1 Moebius 1853 Art. 8. *Distinct points A, B, C, D lie on a circle, C and D on an arc with endpoints A and B, exactly when*

$$m\angle BCA - m\angle BDA = 0, \ i.e., arg(CR(AB, CD)) = 0.$$

Proof Based on the Inscribed Angle Theorem, for a constant k between 0 and π, $\{X : m\angle AXB = k\}$ is one arc with endpoints A and B of a circle on A and B. When the complex number $q = CR(AB, CD)$ is expressed in form $p(cos(\alpha) + \sqrt{-1} \, sin(\alpha))$ for $p > 0$, then $arg \, q = m\angle BCA - m\angle BDA$.

Why? Recall that $CR(AB, CD) = \frac{AC}{BC} \cdot \frac{BD}{AD}$ and $arg \, \frac{AC}{BC} = m\angle BCA$ while $arg \, \frac{BD}{AD} = m\angle ADB$. So C and D lie on a circle which contains A and B, with C and D on an arc with endpoints A and B, exactly when $arg \, q$ equals 0. $\qquad\square$

Theorem 12.2 *When we apply a Moebius transformation to four distinct points lying on a circle, the corresponding four points also lie on a circle.*

Proof Let A, B, C, D be four distinct points lying on a circle, and let A', B', C', D' be the corresponding points under a Moebius transformation. (We assume C and D lie on an arc with endpoints A and B.) Since A, B, C, D lie on a circle, then by the previous lemma, $arg(CR(AB, CD)) = 0$. Since the cross-ratio is invariant under a Moebius transformation, then $arg(CR(A'B', C'D'))$ is also 0. So A', B', C', D' also lie on a circle. $\qquad\square$

We note a logical difficulty when the cross-ratio is defined as in 1853. The claim is clearly true when A, B, C are the three given points on which the *new relationship* is based, but what if $A, B,$ and C are not those original points? Can we assume that the cross-ratios $CR(AB, CD)$ and $CR(A'B', C'D')$ are still equal? Moebius admitted this difficulty in Article 7 of [68] and in the same article indicated how he would avoid it in his 1855 paper [69]. In Article 7 we find

According to this theorem, whose proof I reserve for another opportunity, the new relationship can also be defined by this, that for any four points of one plane which lie on one circle the corresponding points in another [plane] also can be joined by a circle, and thus every circle of one [plane] corresponds to a circle of the other [plane].

In his [69] of 1855, Moebius gave an argument for the preservation of the cross-ratio under what he was then calling the *Kreisverwandtschaft, circle relationship*.

We have already given a proof, in the form presented by Carathéodorie, for the invariance of the cross-ratio. For the interested reader, Moebius's 1855 argument is given in Appendix 6.

Here is another important theorem.

Theorem 12.3 *If a Moebius transformation fixes three distinct points, then it is the identity mapping. More generally, the assignment of values A', B', C' to any three distinct complex numbers A, B, C, respectively, determines a Moebius transformation.*

Proof Consider the number of fixed points of a Moebius transformation, i.e., solutions of $f(z) = \dfrac{az+b}{cz+d} = z$. This is a second degree equation; unless the function is the identity function, there are at most two solutions.   □

3   The Moebius Transform Preserves Angle Measure

Theorem 12.4 *The Moebius transformation preserves angle measure. A mapping is* conformal *when it preserves angle measure.*

Proof *(**Modern**)* Consider the Moebius transformation

$$f(z) = \frac{az+b}{cz+d}, \quad \text{where } ad - bc \neq 0.$$

(i) $f(z) = \dfrac{b - \dfrac{ad}{c}}{c\left(z + \dfrac{d}{c}\right)} + \dfrac{a}{c}$ for $c \neq 0$. Problem: Show $f(z)$ has this form.

This function is the composition of multiplication of the type $g(w) = kw$ for a complex constant k, translation, and the complex function $h(w) = \dfrac{1}{w}$. Translation clearly preserves angle measure. Multiplication by a complex number is the composition of a rotation and a dilation, both of which preserve angle measure.

(ii) Note that $h(w) = \dfrac{1}{w}$ is the composition of the inversion whose center is the origin and radius 1, followed by reflection over the real axis, $y = 0$. Reflection preserves angle measure; that inversion preserves angle measure was proved in the chapter on inversion.

We conclude that the Moebius transformation preserves angle measure.

□

4   Cayley's Matrix Form of the Moebius Transformation

Before we further follow the work of Moebius, it will help to look at a more modern treatment.

Equation (12.1) has a matrix representation, following an 1880 paper of Arthur Cayley [24]. Equation $f(z) = z'$ is represented, where the top row is the numerator of a fraction and the bottom row is the denominator, by

$$\begin{bmatrix} z' \\ 1 \end{bmatrix} = \begin{bmatrix} a & b \\ c & d \end{bmatrix} \begin{bmatrix} z \\ 1 \end{bmatrix}.$$

Assuming $ad - bc \neq 0$, the inverse matrix gives the inverse function, and function composition is carried out by matrix multiplication. (Felix Klein's 1872 paper [58] called on mathematicians to pay attention to the group properties of transformations, including inverse transformations and composition of transformations.)

5   Application 1. Points at Infinity with Moebius Transform

The following example is from Moebius's 1855 proof in [69] that his transformation, mapping plane p to plane p', preserves the cross-ratio. He assumed that his transformation, the *Kreisverwandtschaft*, maps circles and lines to circles and lines. The full proof is in Appendix 6.

Moebius made important use of the idea that one point, M, of the complex plane p was paired with the point at infinity, M', of the second plane, p'. The point of infinity of plane p, denoted N, is paired with point N' of plane p'. A straight line is simply a "circle" which includes the point at infinity.

Moebius also used, without proof, a trait of his transformation which we now refer to as *continuity*. On the one hand, indefinitively close points, *unendlich naher Punckten*, are mapped to indefinitely close points. Further, an indefinitely small triangle, *unendlich klein Drieck*, is mapped to a triangle which both is indefinitely small and is similar. (This claim can be justified in a precise way with the derivative of the function (12.1); Moebius did not think of the function in this algebraic form, so the derviative was not available to him.)

The argument of 1855, in Article 9, is illustrated in our Figs. 12.2 and 12.3. On the left, in plane p, Q and P are indefinitively close to A, on sides AB and AM, respectively, of $\triangle ABM$. M is mapped to infinity, and N, the point at infinity of p, lies on all lines. Points A, Q, B, and N are collinear, so their images, in plane p', lie on a circle, keeping the same order. A, P, M, and N lie, in order, on a line mapped to a line, so A' is between N' and P'. As Q' is indefinitely close to A', then line $Q'A'$ is essentially tangent at A' to circle $A'N'B'$. The vertical angle to $\angle Q'A'P'$ intercepts the arc $A'N'$, as does $\angle N'B'A'$, so those angles are congruent. Angles are part of small triangles and, therefore, are preserved by the Moebius transformation, so $\angle BAM \cong \angle QAP \cong \angle Q'A'P' \cong \angle A'B'N'$.

Fig. 12.2 Moebius Figure of
Article 9, [69], 1855

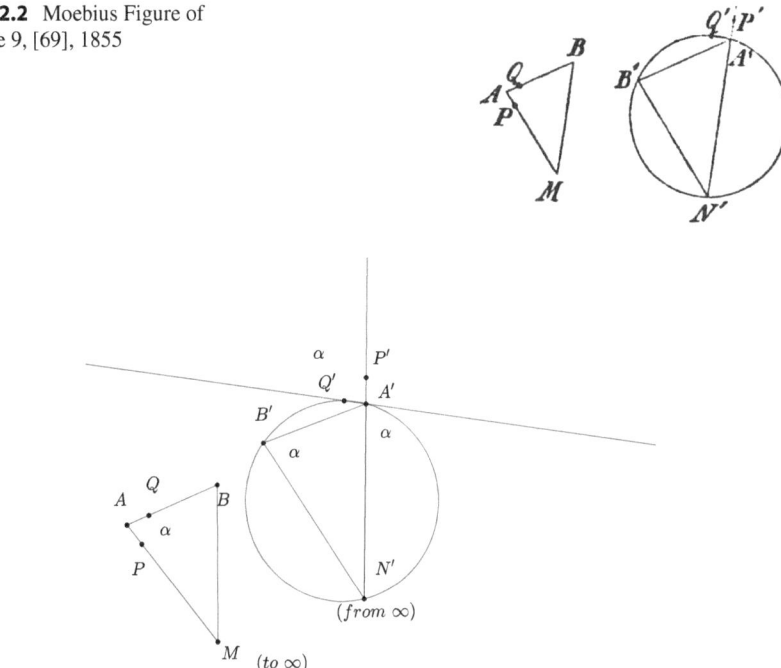

Fig. 12.3 Moebius, based on *Fig.* 9, [62], 1855

Since N is on all three sides of $\triangle ABM$, then N' is the point at which the images of those three sides meet.

If we apply a corresponding argument to two points near to B on sides BA and BM, respectively, we arrive at $\angle ABM \cong \angle B'A'N'$. This means that triangles ABM and $B'A'N'$ are similar, with that correspondence. See Fig. 12.3.

---

6   Application 2. of the Moebius Transformation: An Example from Carathéodorie

Theorem 12.5 *Given two circles, c and d, there is a Moebius transformation mapping them to either a pair of straight lines or to concentric circles. [20, p. 7]*

Proof If the circles meet at a point E, use any Moebius transform that maps E to infinity. This maps the two given circles to lines.

Suppose the circles do not meet. Apply a transformation which sends one circle to a line, a_1, and the other to a circle, b_1. See Fig. 12.4. Let line n be on the center of circle b_1 and perpendicular to a_1, meeting a_1 at M. Draw a tangent line from M to circle b_1, meeting b_1 at T. Draw the circle, d_1, with center M and on point T. Circle

Fig. 12.4 Based on
Carathéodory 1932 p. 7

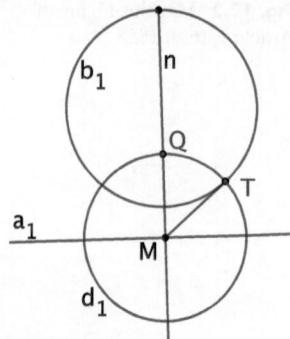

d_1 is orthogonal to circle b_1. Let d_1 meet n at Q. Now take a Moebius transform that sends Q to infinity. Line a_1 and circle b_1 will both be mapped to circles, a_2 and b_2. n will be mapped to a line which is orthogonal to both a_2 and b_2, since n is orthogonal to a_1 and to b_1. Circle d_1 is mapped to a line that is orthogonal to both a_2 and b_2. Since two different lines are orthogonal to circles a_2 and b_2, then a_2 and b_2 must be concentric. $\square$

Note the connection of this theorem to Steiner's Porism. The most crucial step in the proof we offered there shows that two circles which do not meet can be mapped by an inversion to concentric circles. This example from Carathéodorie shows that two circles in the complex plane, which do not meet, can be mapped by a Moebius transformation to concentric circles.

---

7   Exercises—Moebius Transformation

1. (i) Find, in form $f(z) = \dfrac{\alpha z + \beta}{\gamma z + \delta}$, the Moebius transform that maps the circle of radius 1 with its center at the origin to the real axis $y = 0$. (There are different functions that will answer the question. To keep things simple, have f map 1 to 1, -1 to -1, and i to ∞.)
   (ii) Find f^{-1}.
2. Exercise related to Application 1. Suppose that a Moebius transformation maps the point at infinity, Q, to itself.
   (i) Explain why a triangle ABC is mapped to a triangle $A'B'C'$, i.e., why the sides of $\triangle ABC$ are all mapped to lines and not to circles. (Assume Q is not on the triangle.)
   (ii) Why are triangles ABC and $A'B'C'$ similar?
3. Also related to Application 1. When we apply a Moebius transformation to a triangle ABM, with points A and B very near each other, and M mapped to infinity, the image triangle $B'A'N'$ is expected to be nearly similar to $\triangle ABM$.

(*i*) Show this works, approximately, in this example: Let $f(z) = \dfrac{2z - 1}{z + 2}$, and $A = .2 + .1i$, $B = 0$.

Compute A' and B'.

Find M, the point mapped to ∞ by f, and find N', which equals $f(\infty)$.

Compute, by calculator or computer, lengths AM, AB, MB, $B'N'$, $B'A'$ and $N'A'$ then show that $\triangle AMB \sim \triangle B'N'A'$, where the similarity is approximate.

(*ii*) Verify that $\triangle AMB$ and $\triangle B'N'A'$ have opposite orientations, i.e., the walk A to M to B and the walk B' to N' to A' will be one clockwise and one counterclockwise.

4. Let $A = 3 + 4i$, $B = 5$, $C = -4 - 3i$, $D = -5$. Compute $CR(AB, CD)$ and $CR(AC, BD)$. Note that the four points lie on a circle, so the cross-ratios should be real. Can you predict before the computation which will be positive and which will be negative?

5. Describe the Moebius transformation whose fixed points are $1 + 2i$ and ∞ and which maps 2 to 3.

6. Alternative approach to a problem, p. 52, from Hans Schwerdtfeger, *Geometry of Complex Numbers* 1962, University of Toronto Press, reissued by Dover, NY, 1979.

A *perspectivity* is a Moebius transformation of the complex plane in which there are two lines, l and l_1, meeting at Z, and a point S on neither l nor l_1, so that each z of l is mapped to z_1 of l_1 when S, z, and z_1 are collinear.

(*a*) Let us suppose that the two lines meet at finite point Z, and that A is the point of l so $SA \parallel l_1$, and B_1 is the point of l_1 so $SB_1 \parallel l$. Now define the Moebius transformation which carries out the mapping described above, mapping each w to w', in which A, Z, ∞ are mapped, respectively, to ∞, Z, B_1. Suggestion: Use the invariance of the cross-ratio under a Moebius transformation.

(*b*) Use the function rule of the Moebius transformation to show that S is the second fixed point of the transformation, along with Z.

(*c*) Now suppose the lines l and l_1 are parallel, and that S is still a point on neither l nor l_1. Pair each point x of l with x_1 of l_1 when x, x_1, and S are collinear. Define the Moebius transformation which carries out this mapping. Let X, Y of l be paired, respectively, with X_1, Y_1 of l_1. In this case, ∞ will be paired with itself.

(*d*) Prove that the Moebius transformation of part *c*. pairs S with itself.

Topic After 1855: Beltrami-Klein Model

13

As noted in the introduction, we end our main story with 1855. We do this because the geometry practiced by mathematicians underwent a great change once the viewpoint expressed by Riemann in 1854 was absorbed by those mathematicians. Further, an exposition of geometry after 1855 would take us to a level well above that intended for this book; to move beyond 1855 generally takes us to graduate level mathematics, as found in university departments of mathematics and physics.

However, there are two exceptions, both of broad interest outside those university departments, and both involving transformations.

1   The Beltrami-Klein Model

The first exception is an important development in *non-Euclidean* geometry. The term *non-Euclidean* has a particular meaning, generally not including projective geometry. It is geometry based on the postulates of Euclidean geometry except that the Parallel Postulate, Euclid's Fifth Postulate, is replaced by an alternative statement about the existence of parallel lines. For *Hyperbolic Geometry*, the new postulate is that on a point outside a given line there are at least two parallels to the given line. With *Elliptic Geometry* there are no parallel lines. (These names are from Felix Klein, 1871 [59]. See [30].) By 1831, both Janos Bolyai (1802–1860) and Nikolai Ivanovich Lobachevsky (1792–1856) had shown examples of hyperbolic geometries. But they had not demonstated that their geometry includes all those propositions of Euclid's geometry that do not require the Parallel Postulate, while Euclid's Fifth Postulate fails. Such a demonstration would prove the independence of a parallel postulate with respect to the postulates of Euclidean geometry. Said another way, this would guarantee that all efforts to prove Euclid's Fifth Postulate as a theorem within Euclidean geometry were destined to fail.

The first proof of the independence of the Parallel Postulate by means of hyperbolic geometry was based on work by Eugenio Beltrami (1835–1900). In 1868

© The Author(s), under exclusive license to Springer Nature Switzerland AG 2025   147
C. Baltus, *Geometry by Its Transformations*, Compact Textbooks in Mathematics,
https://doi.org/10.1007/978-3-031-72281-3_13

Fig. 13.1 Left: based on
Beltrami figure 1868, p. 294.
Right: Klein's circle model

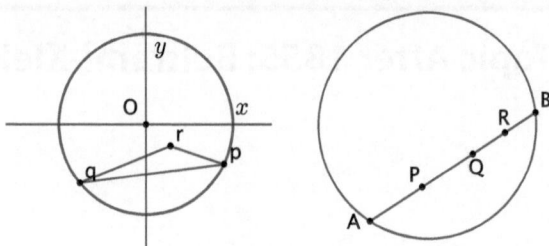

[11], Beltrami presented as a model the interior of a circle, where the "lines" are chords of that circle (without the endpoints). See Fig. 13.1.

Arthur Cayley (1821–1895), in 1859 [23], had discussed distance as an abstract concept, where an essential property was that if Q lies between P and R on a line, then the distance between P and R must equal the sum of the distances from P to Q and Q to R. He proposed a distance formula based on the logarithm.

In 1869 and 1870, the young Felix Klein (1849–1925) learned of the work of Cayley and Beltrami, that of Lobachevsky, and of correspondence of Gauss in which Gauss discussed, approvingly, the work of Lobachevsky. See [1]. Klein used the circle model as proposed by Beltrami and adapted Cayley's distance formula, to define the distance between points P and Q on the chord with endpoints A and B:

$$dist(P, Q) = c \cdot log \left| \frac{AQ \cdot PB}{AP \cdot QB} \right|,$$

where c is a positive constant, and where the points are ordered $A - P - Q - B$.

One can check that as P approaches A or Q approaches B, then $dist(P, Q)$ approaches infinity.

It is an extensive exercise to show that the Beltrami-Klein model satisfies all the postulates of Euclidean geometry, except that it violates Euclid's Fifth Postulate. That collection of postulates would be cleaned up, especially by David Hilbert (1862–1943), to cover assumptions made but not acknowledged by Euclid [56]. We will just note that Euclid's first four postulates are satisfied by the Beltrami-Klein model: any two points determine a line; any line can be extended as much as needed, since the log definition of length gives a chord an infinite length; by that same metric, for any given center and radius there is the set, a "circle," of points at that given distance from the given center. On the other hand, given a line, like pq in Fig. 13.1 Left, and outside point r, the two distinct lines rp and rq are parallels to pq. One of the additional requirements added since the time of Euclid is *betweenness*, that if Q is *between* P and R, then $dist(PR) = dist(PQ) + dist(QR)$. It is an Exercise to check that the log definition of distance satisfies this requirement.

Angle measure can be defined in the Belterami-Klein model, which we will not venture into here. We just note that the perpendicular to a line m on a point Q is found by locating the pole, S, of m with respect to the circle; then the part of line

QS that lies within the circle is the *perpendicular* to m on Q. One can verify that Euclid's Fourth Postulate holds, that all right angles have equal angle measure.

---

2   Exercise—Beltrami-Klein Model

1. Exercise. See Fig. 13.1 Right. We define distance between points P and Q in the interior of a circle by $dist(P, Q) = c \cdot log \left| \dfrac{AQ \cdot PB}{AP \cdot QB} \right|$, when P and Q lie on chord AB, in order $A - P - Q - B$. Prove that when Q lies between P and R, then $dist(PR) = dist(PQ) + dist(QR)$.

Topic After 1855: Isometries and Dilations in French Schoolbooks

14

The transformations we have examined are found in undergraduate mathematics courses, most in College Geometry, while the Moebius transformation is in undergraduate courses in Complex Analysis. An important development occurred, not in higher level mathematics but in French schoolbooks, where geometry moved beyond that of Euclid'd *Elements*. This was the introduction of *geometrical transformations*, not just the dilation, but also the *isometries*: reflection, rotation, translation. (An *isometry* is a transformation in which corresponding distances are equal.)

This movement began with Charles Méray (1835–1911). He had three main goals in the textbook he wrote in 1874 [65] and the reforms he advocated. The first was the amalgamation of plane and spacial geometry into a single subject in the school curriculum. The second was a careful treatment of motion, which we see as the study of isometries in school geometry. The third called for the development of geometry as growing from experience. Thus we see the translation transformation introduced as the type of change that occurs when a desk drawer is moved in and out. And then lines are called *parallel* when one is the image of the other by a translation. He treated translation implicitly as a function, for example, in saying that a sequence of translations can be replaced by a single translation. There follows rotation, beginning with rotation about a line in space; reflection over a line n in a certain plane, for Méray, was a 180° rotation in space about the given line, n.

With the *George Leygues Reform* of 1902–1905, there were significant changes in school mathematics along the lines of Méray's three goals. Several prominent mathematicians were involved, particularly Gaston Darboux (1842–1917), Jacques Hadamard (1865–1963), and Émile Borel (1871–1956).

1 Isometries

Let us start with isometries. Here are relevant modern definitions.

© The Author(s), under exclusive license to Springer Nature Switzerland AG 2025
C. Baltus, *Geometry by Its Transformations*, Compact Textbooks in Mathematics,
https://doi.org/10.1007/978-3-031-72281-3_14

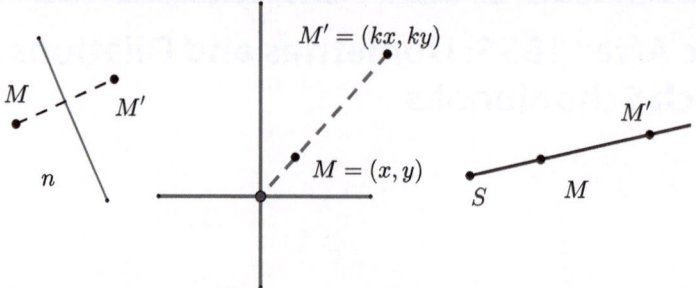

Fig. 14.1 Left: M is mapped to M' in reflection over line n. Center and Right: The dilation with center S and scale factor k maps M to M'

Definitions
An *isometry* is a transformation of the plane (or space) in which distances are preserved.

Translation by vector (a, b) maps each point (x, y) to $(x + a, y + b)$.

Reflection over line n maps each point M to point M' so that n is the perpendicular bisector of segment MM'. (Points of n are mapped to themselves.) See Fig. 14.1 Left. The function that reflects points over line n can be denoted r_n, so we can write $r_n(M) = M'$. Hadamard simply called a reflection a *symmetry*.

For a given planar figure $\mathcal{A}$, a line n is a *line of symmetry*, also called an *axis of symmetry*, if $r_n(\mathcal{A}) = \mathcal{A}$.

Rotation by angle θ about a point S is the mapping of each X to X' where $SX = SX'$ and $m\angle XSX' = \theta$.

(Second definition) *Translation* is the composition of reflections over two parallel lines.

(Second definition) *Rotation* is the composition of reflections over two lines that meet.

All *isometries* are necessarily collineations, mapping lines to lines. It turns out that they are also homologies or compositions of homologies when considered in the projective plane. Recall that a transformation is a homology if it is a collineation with either a point on which all lines are fixed or its dual, a line on which all points are fixed.

Why is a reflection over a line n a homology? It is a collineation since a line is mapped to a line, and n is a line of fixed points. Where is the center? To answer this question, we need to think of a translation as occurring in the projective plane, with a line at infinity.

How about a translation? On the one hand, a translation is the composition of reflections over parallel lines, of twice the length and with the direction of a vector from one line to the other, perpendicular to the lines. (The order of the composition matters.) On the other hand, a translation maps each line to a parallel line, possibly itself, so the line at infinity is a line of fixed points. As a homology, where is the *center* of a translation? (An Exercise asks for the center of a reflection in a line and the center of a translation.)

Rotation about a point, O, is the composition of two reflections over lines which meet at O. Except for the 180 degree rotation, it is not a homology: there is no line of fixed points.

> **Definition** We say that two congruent triangles ABC and abc, with the correspondence of A with a, etc., have *the same orientation* if a walk around the interior of the triangles, in order $A - B - C$ and $a - b - c$, are both clockwise or both counterclockwise. Otherwise, we say they have *reverse* or *opposite orientation*.

Congruent planar figures under a particular correspondence of points will have the orientation of any corresponding pair of triangles within the figures. We note this theorem:

Theorem 14.1 *Reflection over a line in a plane reverses the orientation of figures.*
Since a translation or a rotation in a plane is the composition of two reflections, then translation and rotation maintain the orientation of a figure.

Both Hadamard and Émile Borel wrote textbooks that placed transformations in a group structure. Written in 1905, published in 1910, was Borel's *Géométrie: 1er et 2nd cycles* [15].
He states,

> Geometry is the study of groups of movements. More and more, putting the dynamic study of phenomenon in place of the static study is an essential tendency of the modern spirit. ... I have sought to write a geometry that is more concrete, where the considerations of symmetry, of displacements, are invoked as often as possible. The demonstrations that result from this are simpler and seem to me to be clearer than the Euclidean demonstrations.

In Borel's statement we find all the features that Méray championed in his 1874 text, with special emphasis on movement and a new feature, the discussion of group properties. We will have more to say below.

At least one important treatment of isometries was in German, the *Lehrbuch der Elementar-Geometrie*, by J. Henrici and P. Trautlein, of 1881. The authors trumpeted the introduction of movement into school geometry.

> As opposed to the rigidity and invariance of geometric figures by Euclid, here in the modern viewpoint is the creation of figures by movement and the change of their positions with greater clarity and more natural justification. It is quite superfluous, from a pedagogical

Fig. 14.2 Henrici and
Trautlein, 1881, Figure 71. S
is the center of the rotation
mapping one triangle to the
other

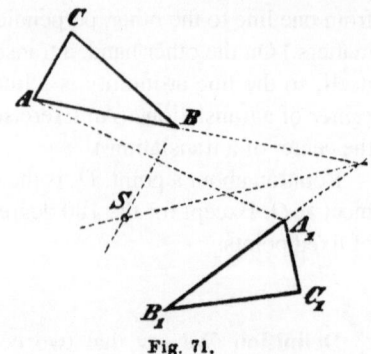

Fig. 71.

regard, to comment here that the teacher, indeed, in handy figures, can let the changes
of position run, until he quickly finds the student able to operate with the concepts that
determine the movement. [55]

In Articles 9 and 10, the perpendicular bisector of a segment is identified as an
axis of symmetry, as is an angle bisector.

In Article 17, arguing that any pair of congruent triangles of the same orientation
are related by a translation or a rotation, Henrici and Trautlein give a construction
of the center of such a rotation. See *Fig.* 71, our Fig. 14.2, where the perpendicular
bisector of segment AA_1 and the bisector of the angle formed by sides AB and
A_1B_1 meet the center S. We ask for a justification in an Exercise.

Triangle congruence is treated in Article 21. With triangles of equal orientation
that agree in Side-Angle-Side, one can be rotated so its sides are parallel to the
corresponding sides of the other, after which the figures coincide by a translation. In
this way the Side-Angle-Side Triangle Congruence Theorem is established. There
is a similar demonstration of the Angle-Side-Angle Theorem.

2   Finite Fixed Points of an Isometry

The various isometries can be identified by their fixed points and whether they keep
or reverse orientation. The result is the following theorem. Completion of the proof
is an Exercise.

Theorem 14.2 (Isometry Theorem)

  (i) *An isometry is determined by the image of three points which are not collinear.
      In particular, if three points not collinear are fixed, then the isometry is the
      identity.*
 (ii) *Angle measure is preserved by an isometry.*
(iii) *If two points are fixed by an isometry, then the line on those points is fixed
      pointwise. The isometry must be a reflection or the identity.*

(iv) *If exactly one point is fixed and orientation preserved, then the isometry is a rotation.*

 (v) *If an isometry preserves orientation and maps a point A to A' and a point B to B', where segments AA' and BB' are not parallel, then the perpendicular bisectors of AA' and BB' meet in a fixed point S.*

(vi) *If a point is fixed and orientation is reversed, then the isometry is a reflection. This means that an isometry which reverses orientation is either a reflection or has no fixed point.*

(vii) *An isometry which preserves orientation and has no (finite) fixed point is a translation.*

(viii) *Every isometry is either the identity, or*
      a reflection, or
      a translation, or
      a rotation, or
      a translation followed by a reflection. (This isometry is called a glide reflection.*)*

 (ix) *Finally, it follows that all isometries are the composition of one, two, or three reflections.*

Proof (Partial) (X' denotes the image of X, etc.) (*i* and *ii*) Given three points not collinear, any other point, Z, is determined by its distance from those three points, for any three circles whose centers are not collinear can meet in at most one point. So Z' is the unique point at the same distances from the images of the three given points. Angle measure is preserved by an isometry since any triangle is mapped to a triangle that is congruent by SSS.

(*iv*) See Fig. 14.3. Let S be the fixed point, and X and Y points not collinear with S. $\triangle SXY$ and $\triangle SX'Y'$ are congruent, with the same orientation, so $\angle XSY \cong \angle X'SY'$. It follows that $SX = SX'$, $SY = SY'$, and $\angle XSX' \cong \angle YSY'$. Thus, X and Y are rotated by the same angle about S, $\angle XSX'$. This isometry agrees on three points with the noted rotation, so the isometry is that rotation.

(*vi*) Let S be a fixed point when the orientation is to be reversed. Draw a triangle XYS, where X and Y are not collinear with S, and draw the image triangle $X'Y'S$. Show the bisector of $\angle XSX'$ is the bisector of $\angle YSY'$, so that bisector is the perpendicular bisector of both segment XX' and of segment YY'. This means that

Fig. 14.3 Isometry
Theorem, part *iv*

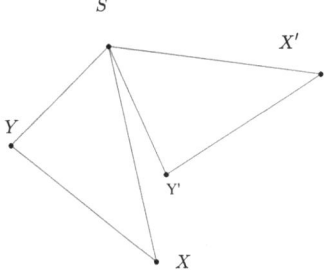

the entire plane is reflected over that angle bisector, since the isometry acts as a reflection on three points.

(vii) Given an isometry with no finite fixed points that preserves orientation, let ABC be a triangle whose image is the congruent $\triangle A'B'C'$, and where no side is of $\triangle ABC$ is parallel to line AA'. By (v), lines AA', BB', CC' are mutually parallel. Now, since side $A'B'$ is congruent to side AB, there are, in general, two possible positions for point B', but so that $\triangle A'B'C'$ has the orientation of $\triangle ABC$, B' is the point so $ABB'A'$ is a parallogram. Likewise, $ACC'A'$ is a parallelogram. So segments AA', BB', CC' are mutually parallel and congruent. So the isometry is a translation.

($viii$) The above cases cover all the possibilities where there is a fixed point. When there is no fixed point, the isometry may be a translation or else it reverses orientation. See Exercise 10, which asks for a proof that when orientation is reversed and there are no fixed points, then the isometry is the composition of a translation followed by reflection.   □

3   The Context in Which Transformations Appeared in School Mathematics

What is the historical context in which geometric transformations appeared in school mathematics?

Movement As Rudolf Bkouche observed [14, p. 187], the Greeks tried to eliminate movement from their mathematics for philosophical reasons. Geometry was to be about the *essence* of a figure, which is static. They developed some curves by a movement, but those belonged more to physics. The Greek attitude lost its hold in the early modern period as scientists such as Galileo and Newton gave precise mathematical descriptions of the motion of bodies.

The introduction of "movement," in the form of isometries, was a key part of France's new mathematics school curriculum in the *La Reform Georges Leygues* of 1902, and its implementation by 1905. The program redesign of the mathematics curriculum was overseen by a prominent mathematician, Gaston Darboux. In an address of 1914, Darboux listed four essentials of the reforms of 1902–1905. The second was

> The systematic employment, in geometry, of methods of transformation which simplify the study, and bring with it a principle of classification. [72, p. 5]

The change appeared in instructions from the Ministry of Education. The *Bulletin administratif du ministère de l'instruction publique*, 1905, included under "Instruction relative a l'enseignement des mathématiques, 1er cycle B"

> A constant appeal to movement seems to facilitate the teaching of geometry: in this way, parallelism will be tied to the experimental notion of translation, the study of perpendicular

planes and lines will be tied to rotation, the idea of equality will be tied to the transport of figures which one will make precise in introducing the simple notion of orientation. [66, p. 707]

Pedagogy J. H. Pestalozzi (1746–1827), building on romantics like Jean Jacques Rousseau, put into practice the idea that learning should be a passage from the concrete and familiar to the abstract.

We've seen, for example, that Méray introduced *translation* as like the movement of a drawer that is well set in its frame. Developing geometry from transformations was taken as more intuitive, and thus more convincing. Consider the Isosceles Triangle Theorem, that base angles of an isosceles triangle are congruent. Reflecting an isosceles triangle over the bisector of its vertex angle maps one base angle onto the other, showing those angles are congruent. This seemed clearer than the elaborate proof in Euclid's Book 1 Prop. 5.

Group Concepts We note, finally, Felix Klein's *Erlanger Program*, announced in an address of 1872 and slowly accepted. The first published English translation appeared in 1892. Quoting from that translation,

> geometric properties are characterized by their remaining invariant under the transformations of the principle group. [58]

Already in Poncelet's *Traité* of 1822, we see importance attached to features invariant under transformations. "A figure whose parts have relations indestructible by the effects of a projection will be called . . . a *projective figure*." [Art. 2, 3, 4]

Through the nineteenth century, implicit concepts of group theory, especially with permutations, became more prominent. (See Wussing, *The Genesis of the Abstract Group Concept*, 1969, 1984.)

Group theory concepts appeared in secondary school mathematics after 1900. Hadamard, late in his text of 1906, included a short introduction to groups.

> 291. Groups. The property of possessing invariants is, for the transformations of which we have just spoken, the consequence of another fundamental property about which we will say a few words.
> One gives the name *product* of two or several transformations, to the transformation which equals those effected successively in the order in which they are named.
> Example. – The result of Art. 102 can be enunciated thus:
> *The product of two symmetries [reflections] is a rotation or a translation.*
> With that, one says that a certain set of transformations constitutes a *group* if the product of any two among them is still a transformation belonging to the set.
> Thus *the set of all dilations (homothéties) is a group*: this reduces to saying that two figures homothétiques to a third are homothétiques between themselves. The set of all the dilations whose poles (centers) lie on a certain straight line form a group, since we know that the center of the dilation which is a product of two others is on a line with the centers of those.

This same viewpoint appeared late in the *New Mathematics* movement. Irving Adler, in 1968, listed *Goals of High School Geometry*, which included "2. Introduction to the role of transformations of space in the study of geometry." He was referring to isometries: translation, rotation, reflection, and glide reflection, including "multiplication" of isometries.

> the study of congruent figures in Euclidean plane geometry is part of the study of the geometry associated with the group of *isometries* of the plane, where an isometry is a transformation of the plane that preserves the distances between points. [2, p. 228].

4   Similarity and Congruence by Geometric Transformations

Hadamard, as in the earlier quotation from 1906, treated dilations, which he called *homothéties*, alongside isometries. The integration of the two topics was thorough in the few books that developed what came to be called *transformation geometry* in the 1970s.

An example of a *transformation geometry* text is [31] of 1971, with a variation offered by a major publisher until 1993. The congruence and similarity of planar figures can be defined in terms of isometries.

We first remind readers of the two alternate definitions of the dilation with *scale factor k*. See Fig. 14.1 Center and Right.

Definition
First definition: the *dilation* with non-zero *scale factor k* and *center* at the origin, maps point (x, y) to point (kx, ky).
Second definition: the *dilation* with non-zero *scale factor k* and *center* at point S maps a point M to point M' on line SM where ratio SM'/SM equals k. The lengths in the ratio are signed. (S is mapped to itself.)
     Here are definitions in the 1991 text [32].

Definition
Two figures F and G are *congruent figures*., . . , if and only if G is the image of F under a reflection or composite of reflections.
     . . .
A transformation is a *similarity transformation* if and only if it is the composite of size changes [dilations] and reflections.
Two figures F and G are *similar*, . . . , if and only if there is a similarity transformation mapping one onto the other. p. 279, 586

As in the quotes above, similar figures can be introduced as figures related by a dilation and isometry. The triangle similarity theorems: SSS, AA, and SAS, can be derived by applying a dilation to one triangle to make it congruent to the other given triangle. An example and an exercise are in the section of Exercises.

5 Exercises—Isometries

1. Suppose we have a rotation by angle θ and with center S that maps line m to line m_1, mapping points A and B on m to points A_1 and B_1, respectively, on m_1. Let lines m and m_1 meet at X.
   (a) Prove that center S lies on the perpendicular bisector of segment AA_1.
   (b) Prove that center S lies on the bisector of angle AXA_1.
   Parts (a) and (b) together prove that S, the center of the rotation of $Fig.$ 71 of the Henrici and Trautline text is the intersection of the perpendicular bisector of segment AA_1 and the bisector of the angle at which lines AB and A_1B_1 meet. Note: By the Isometry Theorem, an isometry which preserves orientation is a translation or a rotation. The triangles in the Henrici and Trautline $Fig.$ 71 are not related by a translation, so there is a rotation mapping one to the other.
2. Complete the proof of Parts (iii), (v), (vi), and (ix) of the Isometry Theorem.
3. Here is a proof involving dilations of the SSS triangle similarity theorem.
   Given: $\triangle ABC$ and $\triangle DEF$, where

   $$\frac{DE}{AB} = \frac{DF}{AC} = \frac{EF}{BC}.$$

   Show: $\triangle ABC \sim \triangle DEF$.
   Our plan is to apply a dilation to $\triangle ABC$ to make its image congruent to $\triangle DEF$. Let $\frac{DE}{AB}$ be the positive constant k. Apply to $\triangle ABC$ the dilation with center A and scale factor k. Since all lengths are multiplied by the scale factor, the side lengths of the image triangle $A'B'C'$ are $A'B' = k \cdot AB = DE$, $A'C' = k \cdot AC = DF$, $B'C' = k \cdot BC = EF$. By the SSS triangle congruence theorem, $\triangle A'B'C' \cong \triangle DEF$. It follows by the definition of similarity that $\triangle ABC \sim \triangle DEF$. $\square$
   Exercise. Use dilation to prove the AA and SAS triangle similarity theorems.
4. Use isometries to prove that two triangles which agree in $Angle - Side - Angle$ are congruent. There are two cases: when the triangles have the same orientation and when they have opposite orientation. Suggestion: Use either a rotation or a reflection as the initial step.
5. Find the 2-by-2 matrix which reflects points of the plane over line $x + y = 0$.
6. Find the image of $A = (3, -1)$ and $B = (4, 1)$ when rotated by 135 degrees about the origin. (Leave $\sqrt{2}$ in the answer.)
7. Consider reflection over a line and translation as homologies when the real plane is expanded to the projective plane, with the line at infinity. As noted before, for

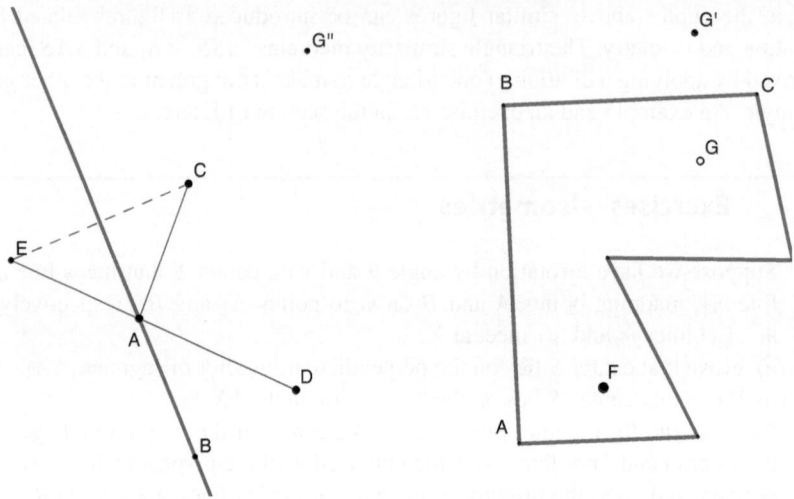

Fig. 14.4 Left: Path of reflection from C to D. Right: Miniature Golf Exercise

reflection over a line m, m is the axis, and for translation the line at infinity is the axis.

(a) For reflection in a plane over a particular line, m, what is the *center*?
(b) For translation in a plane by a particular vector x what is the *center*?

8. We know that if lines m and n meet at S, then reflection over line m followed by reflection over line n is a rotation about S. What is the angle measure of the rotation?

9. Describe in simple and precise language the collineation with center $(2, 3)$, where the line at infinity is a line of fixed points, and which maps point $(3, 2)$ to point $(0, 5)$.

10. Referring to the Isometry Theorem, prove that an isometry which reverses orientation and has no fixed point can be achieved by a reflection over a line followed by a rotation. Explain how the isometry can map a triangle XYZ to triangle $X'Y'Z'$.

11. Miniature golf.

Reflection over a line can guide a player of billiards or of miniature golf who wishes to bounce a ball off a wall. "When a ball is rolled without spin against a wall, it bounces off the wall as if it had gone through the wall and its path were reflected over the wall. The ball takes the shortest path to its destination." [32, p. 273] The same principle applies to a ray of light reflecting off a wall. See Fig. 14.4 Left. There point C is reflected over line AB to point E. By a property of reflection, a path from C to point A on line AB and then to D has the same length as the path from E to A to D. If A is allowed to vary along line AB, the shortest trip from E to A to D occurs when A is on line ED. A proof, unlikely to be original, of this *shortest path* property is found in the *Catoptrica* by Hero of Alexandria, from, it appears, the first century AD. (That nature "chooses" to expend the least effort in various situations is known by several names, including *Fermat's Principle* or the *Principle of Least Time*.)

To plot the path of a golf ball from point F to the hole at G, reflect G to G' over side BC, and then reflect G' to G'' over side AB. The golfer's shot must strike side AB where it meets line FG''. (One also could reflect F over line AB, then connect A' to G'.) Plot the path of the golf ball. Then create a new miniature golf problem.

1. Given two circles of equal radius, meeting at A and B, where the center of each lies on the other. Draw a line on A meeting one circle at C and the other at D. Prove that $\triangle BCD$ is equilateral.
2. Show that the composition of two linear functions of form $f(x) = ax + b$, f_1 and f_2, is itself a linear function.
3. Prove the converse of the Cyclic Quadrilateral Theorem. In other words, if a pair of opposite angles of quadrilateral $ABCD$ are supplementary, then A, B, C, and D lie on a circle.
4. Pivot Theorem. Given triangle ABC and points A', B', C' on the sides, A' on the side opposite vertex A, B' opposite B, and C' opposite C. Draw three circles: one on A', B' and C, one on B', C' and A, and one on C', A' and B. Then those circles meet at a single point Q.
Hint: Let Q be the point, other than B', where the first two circles meet. Quadrilaterals $A'CB'Q$ and $B'AC'Q$ are cyclic. Show that quadrilateral $BA'QC'$ is also cyclic. (It follows that Q lies on the circle on B, A', and C'.)
5. Forder p. 20 [45]. Given points A', B, D, and C on a circle, point A not on this circle, point B' on circle DAC, point C' on circle DAB. See Fig. A.1. Show that circles $AB'C'$, $CB'A'$, and $BC'A'$ all have a common point. That point would be D' in the figure.
A proof may be found by performing an inversion. Point D could serve as center for the inversion. Exercise 4 may help.
6. In a circle, two quadrilaterals are inscribed, one with sides a, b, c, d, the other with sides a_1, b_1, c_1, d_1, where $a \parallel a_1, b \parallel b_1$, and $c \parallel c_1$. Prove $d \parallel d_1$.
Note. Poncelet, in 1820 Art 73, had this problem: If one inscribes in a circle, as you like, a quadrilateral whose first three sides lie on three given collinear points, then the fourth side must always lie on a fourth fixed point that is collinear with the other three. If we project so the line goes to the line at infinity and the circle to another circle, we have this problem.
7. (From I. M. Yaglom.) Given a circle with points A and B on the circle,

Fig. A.1 H. G. Forder,
Geometry, 1960, p. 21

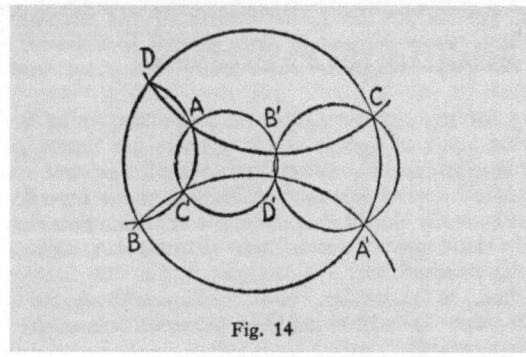

Fig. 14

Fig. A.2 Problem from
Yaglom, *Geometric
Transformations* IV, p. 14

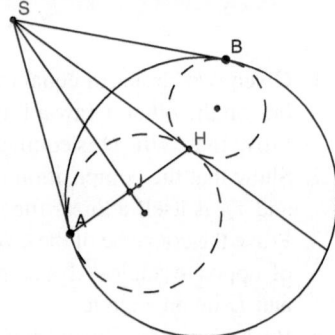

(*i*) Construct a pair of circles, one tangent to the given circle at A and the other tangent at B, so the constructed circles are tangent to each other, and

(*ii*) describe the path of points H at which the two constructed circles are tangent to each other. See Fig. A.2 [107, p. 14].

8. Let S, A, A_1, and C be collinear points on a line m and L a point not on m. There is a homology with center S, mapping A to A_1, and mapping C to infinity, where L is a fixed point. Find the axis of this homology.

   Hint. Find the image of line LC.

9. Let c be a circle and circle c' its image under Inversion with center S. The inversion maps point A to A' and maps point B to B' as in Fig. A.3, with S, A, A', B, B' collinear. (Figure A.3 is the case that the circle of inversion meets circle c, but the problem also refers to the case where the circle of inversion does not meet circle c.)

   Any two circles are related by a dilation. Describe the dilation by naming its center and expressing its *scale factor* in terms of j, the radius of the circle of inversion, and the power, p, of S with respect to circle c.

10. Desargues' Theorem and parallel lines.

    (a) What can we conclude by Desargues' Theorem in the case of triangles ABC and $A_1B_1C_1$ in perspective from a point S when sides AB and A_1B_1 are parallel?

    (b) Prove the following theorem:

Fig. A.3 Inversion related to
a dilation

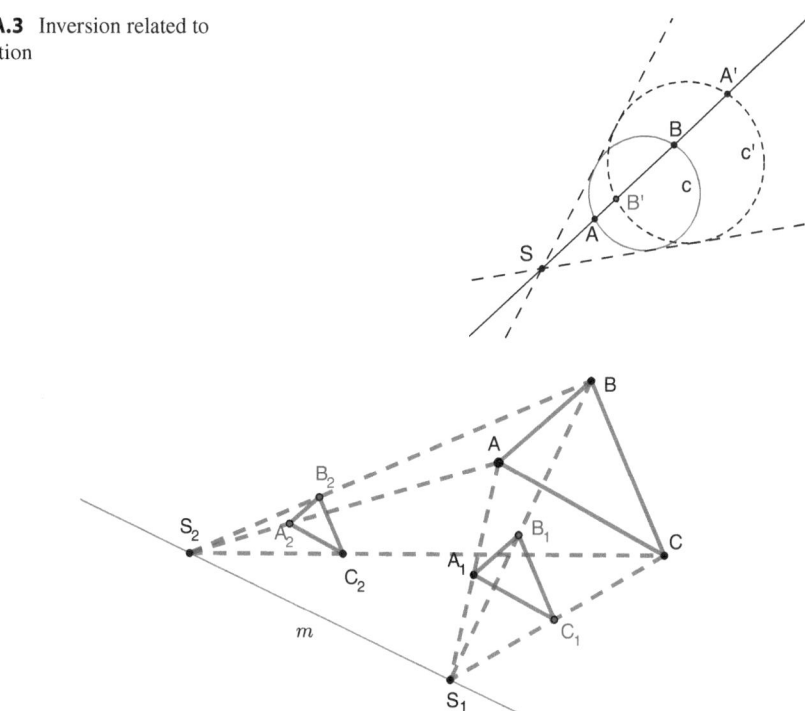

Fig. A.4 Exercise 10*b*. Composition of dilations with centers S_2 and S_1 on line m

Theorem A.1 *The composition of a dilation with center S_1 with a dilation with center S_2 is, itself, a dilation whose center is on line $S_1 S_2$.*

In Fig. A.4, we have the case of the dilation with center S_1 that maps point A_1 to A, applied to $\triangle A_1 B_1 C_1$ to produce $\triangle ABC$, and then the dilation with center S_2 that maps A to A_2, applied to $\triangle ABC$ to produce $\triangle A_2 B_2 C_2$. Prove that in this case, triangles $A_1 B_1 C_1$ and $A_2 B_2 C_2$ are related by a dilation whose center is collinear with S_1 and S_2.

Suggestion Apply Desargues' Theorem twice, once to triangles $A_1 B_1 C_1$ and $A_2 B_2 C_2$, and a second time to triangles $S_1 B_1 C_1$ and $S_2 B_2 C_2$.

11. An alternate proof of Pascal's Theorem. The proof and figure are from Dowling [37, p. 68] (Fig. A.5).

Theorem A.2 *Pascal's Hexagon Theorem. If a hexagon is inscribed in a conic section, then the opposite sides meet in collinear points.*

Fig. A.5 Proof of Pascal
Theorem, Dowling Fig. 38

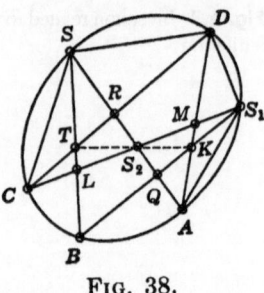

FIG. 38.

Let A, B, C, D, S, S_1 be points on a given conic. We join these points as in
Dowling's *Fig*. 38, our Fig. A.5. Because the points are on a conic, pairing $S_1 D$
with SD, $S_1 A$ with SA, $S_1 B$ with SB, etc, is a projective pairing of the pencil on S
with the pencil on S_1. Then we pair the lines of the pencil on S with points on line
DC, and the lines of the pencil on S_1 with points of line DA. Explain why lines
DA and DC are related in a perspectivity, why S_2 is the center of that perspectivity,
and why it then follows that opposite sides of hexagon SBS_1CDA meet in collinear
points.

12. Hadamard p. 213 Art. 218. Suppose that by an inversion with center I point A
    is mapped to point A' and B is mapped to point B'. Show quadrilateral $AA'B'B$
    is cyclic. You may assume that quadrilateral $AA'B'B$ is simple, i.e., sides only
    meet at the four vertices.

13. Note that in homogeneous coordinates, the hyperbola $\dfrac{x^2}{a^2} - \dfrac{y^2}{b^2} = 1$ is
    $\dfrac{x^2}{a^2} - \dfrac{y^2}{b^2} = z^2$.
    (*i*) Write, in homogeneous coordinates (x, y, z) the equation of the unit circle
       with center at the origin.
    (*ii*) Show that the two points $(1, \pm i, 0)$ lie on the unit circle with center at the
       origin. In this sense, the line at infinity meets the circle in two imaginary
       points.
    (*iii*) Find, in homogeneous coordinates, the two points of the line at infinity
       that meet the hyperbola given above. (These points will correspond to the
       asymptotes of the hyperbola.)

14. Given a line l, a point X on l and a circle that does not meet l, construct a circle
    tangent to l at X, and tangent to the given circle.
    Hint: The Common Secant Theorem. Or an inversion can create what seems
    like a simpler problem: to find a circle tangent at a given point on a given circle
    and tangent to a given line.

A Matrix Algebra Primer

B

We will deal with three types of objects. First, a *vector* is an ordered pair or triple of real numbers. We will work in $\mathbb{R}^2$, the real plane, formed of real ordered pairs, or, in $\mathbb{R}^3$, the set of ordered triples (a, b, c). We will use the vector notation, $\vec{v}$, to denote a vector. When it is clear that a certain letter indicates a vector, and not a single number, the arrow may be omitted. In the case of $\vec{v} = (a, b, c)$, we will speak of a, for example, as the *first component* or *first coordinate* of vector $\vec{v}$. Vectors with the same number of components may be added in the way expected, adding corresponding components to produce a vector of the same number of components.

Second, a *scalar* refers to a single real number, although on some occasions a *scalar* may be a complex number. Scalars can be added and multiplied as with any real numbers. We have *scalar multiplication* in which a scalar multiplies a vector. If k is a scalar and $\vec{v}$ the vector (a, b, c) then $k\vec{v}$ or $\vec{v}k$ is the vector (ka, kb, kc).

For $\mathbb{R}^2$, the *basis* will be the *standard basis* $\{\vec{e}_1, \vec{e}_2\}$ where $\vec{e}_1 = (1, 0)$ and $\vec{e}_2 = (0, 1)$. In $\mathbb{R}^3$, $\vec{e}_1 = (1, 0, 0)$, $\vec{e}_2 = (0, 1, 0)$, and $\vec{e}_3 = (0, 0, 1)$.

This means that in $\mathbb{R}^2$, for example, (a, b) is the same as $a\vec{e}_1 + b\vec{e}_2$.

Third, a *matrix* is a rectangular array of numbers. "Rectangular" means that the matrix is formed of rows and columns with the same number of entries in each row and the same number in each column. If M denotes a matrix, then $a_{2\,3}$ denotes the entry in Row 2 and Column 3. If M has 4 rows and 5 columns, we say that M is a "4-by-5" matrix. When dealing with matrices, a vector will be a "row vector"—represented by a matrix of one row—or a "column vector"—represented by a matrix of one column.

Operations Defined

Two vectors of the same number of entries have a *dot product* or *vector product*, carried out by multiplying entry by entry and adding the results. So $(a, b, c) \cdot (x, y, z)$ is $ax + by + cz$. The operation symbol is a dot at midlevel; the dot itself, between

© The Author(s), under exclusive license to Springer Nature Switzerland AG 2025
C. Baltus, *Geometry by Its Transformations*, Compact Textbooks in Mathematics, https://doi.org/10.1007/978-3-031-72281-3

two vectors, can be omitted when the operation is understood to be the dot product. Note that the dot product of two vectors is a scalar, not a vector.

A 3-by-3 matrix, M, multiplies, on the left, a column vector of three entries to produce a column vector of three entries. If the column vector is $\vec{v}$ and the rows of the matrix M are $\vec{r}_1, \vec{r}_2$, and $\vec{r}_3$, then the product is the column vector whose entries are $\vec{r}_1 \cdot \vec{v}$, $\vec{r}_2 \cdot \vec{v}$, and $\vec{r}_3 \cdot \vec{v}$. For example, with $M = \begin{bmatrix} 3 & 2 & 4 \\ 0 & -2 & 2 \\ 1 & 3 & -2 \end{bmatrix}$, and $\vec{v} = \begin{bmatrix} x \\ 3 \\ 5 \end{bmatrix}$,

then $M\vec{v} = \begin{bmatrix} 3 & 2 & 4 \\ 0 & -2 & 2 \\ 1 & 3 & -2 \end{bmatrix} \begin{bmatrix} x \\ 3 \\ 5 \end{bmatrix} = \begin{bmatrix} 3x + 6 + 20 \\ -6 + 10 \\ x + 9 - 10 \end{bmatrix} = \begin{bmatrix} 3x + 26 \\ 4 \\ x - 1 \end{bmatrix}$.

For the product of two 3-by-3 matrices, let the first matrix be M and the three columns of the second be $\vec{c}_1, \vec{c}_2$, and $\vec{c}_1$. The product is a 3-by-3 matrix whose columns are $M\vec{c}_1$, $M\vec{c}_2$, and $M\vec{c}_3$. For example let M be as above and N the

matrix $\begin{bmatrix} 5 & 0 & -1 \\ 2 & -2 & 4 \\ 3 & 1 & -3 \end{bmatrix}$. Then $MN = \begin{bmatrix} 3 & 2 & 4 \\ 0 & -2 & 2 \\ 1 & 3 & -2 \end{bmatrix} \begin{bmatrix} 5 & 0 & -1 \\ 2 & -2 & 4 \\ 3 & 1 & -3 \end{bmatrix} = \begin{bmatrix} 31 & 0 & -7 \\ 2 & 6 & -14 \\ 5 & -8 & 17 \end{bmatrix}$.

Matrices can execute transformations which are *linear*.

Definition
When $\vec{a}$ and $\vec{b}$ are *vectors* and α and β are *scalars*—which for us are real or complex numbers—then a transformation f is *linear* if $f(\alpha\vec{a} + \beta\vec{b}) = \alpha f(\vec{a}) + \beta f(\vec{b})$. (The vectors may be replaced by matrices of the appropriate size.)

By the way matrix multiplication is defined, for a matrix M, a scalar α, and a column vector of correct size, $M(\alpha\vec{x}) = \alpha M\vec{x}$. And when M is a matrix and $\vec{x}$ and $\vec{y}$ column vectors of the correct dimension, then $M(\alpha\vec{x} + \beta\vec{y}) = \alpha M\vec{x} + \beta M\vec{y}$, and if N_1 and N_2 are matrices of the correct size, then $M(\alpha N_1) = \alpha M N_1$ and $M(N_1 + N_2) = MN_1 + MN_2$. Thus, a function on vectors or matrices defined by matrix multiplication must be linear.

On the other hand, suppose we have a linear function f defined on vectors of $\mathbb{R}^2$ or $\mathbb{R}^3$. Say $\mathbb{R}^3$. We can create a matrix M so $f(\vec{x})$ is the matrix product $M\vec{x}$ when $\vec{x}$ is written as a column vector.

How? Let us work in $\mathbb{R}^3$. Let $\vec{c}_1$ be $f(\vec{e}_1)$; let $\vec{c}_2$ be $f(\vec{e}_2)$; and $\vec{c}_3$ be $f(\vec{e}_3)$. Now let M be the 3-by-3 matrix whose columns are $\vec{c}_1, \vec{c}_2$, and $\vec{c}_3$. Then what is $f(\vec{x})$? It is the matrix product $M\vec{x}$ when $\vec{x}$ is written as a column vector.

Why? Let basis vectors $\vec{e}_1, \vec{e}_2$ and $\vec{e}_3$ be written as column vectors. If $\vec{x} = (a, b, c)$, then $\vec{x} = a\vec{e}_1 + b\vec{e}_2 + c\vec{e}_3$. Since f is linear, $f(\vec{x}) = f(a\vec{e}_1 + b\vec{e}_2 + c\vec{e}_3) = af(\vec{e}_1) + bf(\vec{e}_2) + cf(\vec{e}_3) = a\vec{c}_1 + b\vec{c}_2 + c\vec{c}_3$. This last expression is $M\vec{x}$. $\square$

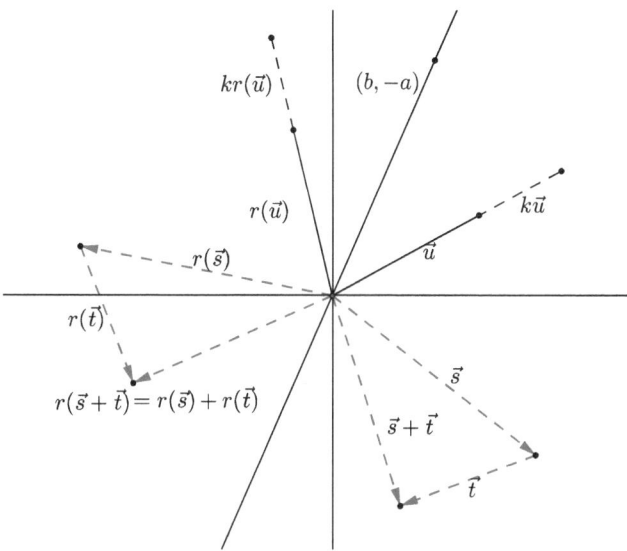

Fig. B.1 Reflection over line $bx + ay = 0$, on the origin, is linear

We have a corresponding rule for 2-by-2 matrices. Let us consider reflection, r, over a line m on the origin in $\mathbb{R}^2$, $ax + by = 0$ where a and b are not both 0. Vector $(b, -a)$ lies on the line. See Fig. B.1. If we consider a vector $\vec{u}$ and a scalar multiple $k\vec{u}$ and their reflections over line m, we see that $r(k\vec{u}) = kr(\vec{u})$. Likewise, reflecting the vector sum $\vec{s} + \vec{t}$ shows that $r(\vec{s} + \vec{t}) = r\vec{s} + r\vec{t}$.

We conclude that reflection over a line on the origin is a linear transformation.

Examples: Finding Matrices for Linear Functions
We can find the 2-by-2 matrix, N, for the *reflection* by giving two points and their images, provided the points are not collinear with the origin. One can find the matrix in terms of a and b, but let us examine the process in the simple case where line m is $-3x + 2y = 0$. Here $(b, -a) = (2, 3)$. Now point $(2, 3)$ is on the line so is mapped to itself. Since $2(3) + 3(-2) = 0$, then vector $(3, -2)$ is perpendicular to line m and, so, is mapped to its negative, $(-3, 2)$.

We make $N = \begin{bmatrix} d & e \\ f & g \end{bmatrix}$, and solve for d, e, f, g in the equations $N \begin{bmatrix} 2 \\ 3 \end{bmatrix} = \begin{bmatrix} 2 \\ 3 \end{bmatrix}$ and $N \begin{bmatrix} 3 \\ -2 \end{bmatrix} = \begin{bmatrix} -3 \\ 2 \end{bmatrix}$. We get $N = \dfrac{1}{13} \begin{bmatrix} -5 & 12 \\ 12 & 5 \end{bmatrix}$.

Now consider *rotation*, f, about the origin by angle θ in the positive direction. Rotation by θ is the composition, in the correct order, of two reflections over lines which meet at the center of rotation in angle $\theta/2$. Composition of two linear functions is linear so rotation about the origin is a linear operation. So it is carried out by matrix multiplication.

We will build the matrix by finding $f(\vec{e}_1)$ as the first column of the matrix and $f(\vec{e}_2)$ as the second column of the matrix. e_1 is rotated to $(cos\ \theta, sin\ \theta)$ and e_2 is rotated to $(-sin\ \theta, cos\ \theta)$, so a vector $\begin{bmatrix} x \\ y \end{bmatrix}$ is mapped to $\begin{bmatrix} cos\ \theta & -sin\ \theta \\ sin\ \theta & cos\ \theta \end{bmatrix} \begin{bmatrix} x \\ y \end{bmatrix}$.

Problem 1 Find the 2-by-2 matrix which maps point $(1, 2)$ to $(2, 2)$ and maps $(-1, 3)$ to $(2, 0)$.

Solution Let $\begin{bmatrix} a & b \\ c & d \end{bmatrix}$ be the matrix M that we seek.

Now, $M \begin{bmatrix} 1 \\ 2 \end{bmatrix} = \begin{bmatrix} 2 \\ 2 \end{bmatrix}$, and $M \begin{bmatrix} -1 \\ 3 \end{bmatrix} = \begin{bmatrix} 2 \\ 0 \end{bmatrix}$ so we have the equations

$$a + 2b = 2, \quad c + 2d = 2, \quad -a + 3b = 2, \text{ and} - c + 3d = 0.$$

We find

$$\begin{bmatrix} a & b \\ c & d \end{bmatrix} = \frac{1}{5} \begin{bmatrix} 2 & 4 \\ 6 & 2 \end{bmatrix}.$$

Group Properties: Identity and Inverse Matrices

Now we consider only square matrices, which we limit to 2-by-2 and 3-by-3 matrices. We consider a collection of elements, such as rational numbers or square matrices of a particular size, with an operation, $\circ$, such as rational number multiplication or matrix multiplication, where there is exactly one solution X within the collection for every equation of form $A \circ X = B$.

For rational numbers, such an equation could be

$$\frac{2}{3} \cdot X = 5.$$

For 2-by-2 matrices, such an equation could be

$$\begin{bmatrix} 1 & -3 \\ 2 & -1 \end{bmatrix} \begin{bmatrix} w & x \\ y & z \end{bmatrix} = \begin{bmatrix} 4 & -3 \\ 1 & 0 \end{bmatrix}.$$

To solve for X in the rational number equation, we multiply on the left, on both sides of the equation, by the multiplicative inverse of $\frac{2}{3}$, namely $\frac{3}{2}$. This gives

$$\frac{3}{2}(\frac{2}{3} \cdot X) = \frac{3}{2} \cdot 5.$$

(We pay attention to the side on which we multiply since, although rational number multiplication is commutative, other operations, such as matrix multiplication, are not.)

We move the parentheses so the left side is $(\frac{3}{2} \cdot \frac{2}{3}) \cdot X$. Now, $\frac{3}{2} \cdot \frac{2}{3}$ is the product of a fraction and its multiplicative inverse, so that product is 1. So our equation is $1 \cdot X = \frac{3}{2} \cdot 5 = \frac{15}{2}$. Since $1 \cdot X = X$, then we have the unique solution $X = \frac{15}{2}$.

© The Author(s), under exclusive license to Springer Nature Switzerland AG 2025
C. Baltus, *Geometry by Its Transformations*, Compact Textbooks in Mathematics,
https://doi.org/10.1007/978-3-031-72281-3

The properties which guarantee a unique solution of every equation of the correct form are the properties of a *group*.

Definition

A set S of *elements* $a, b, c, \dots$ with an *operation* $\circ$ forms a group G (G is the pair S and $\circ$) if

(i) G is *closed*, i.e., $a \circ b$ is defined and in S for any elements a and b in S,

(ii) G is *associative*, i.e., $(a \circ b) \circ c = a \circ (b \circ c)$ for any elements a, b, and c in S,

(iii) G has an *identity element, e*, i.e., there exists an element e of S so $a \circ e = e \circ a = a$ for any element a in S, and

(iv) every element a of S has an *inverse element, a^{-1}*, i.e., for every a in S, there exists an element a^{-1} of S so $a \circ a^{-1} = a^{-1} \circ a = e$, where e is the identity.

Systems of equations of the form $\begin{bmatrix} a & b \\ c & d \end{bmatrix} \begin{bmatrix} x \\ y \end{bmatrix} = \begin{bmatrix} s \\ t \end{bmatrix}$, or the corresponding 3-by-3 system, had been solved for centuries, in Europe and China and Japan, often without explicitly using matrices. What we now call the *determinant* was discovered at various times to answer the question whether a system of n linear equations in n unknowns has a solution.

The system we gave just above is

$$ax + by = s \tag{C.1}$$

$$cx + dy = t. \tag{C.2}$$

When we multiply the top equation by c and the lower equation by a and then subtract, we reach the equation $y(ad - bc) = af - ce$. There will be a unique solution when $ad - bc$ is not 0.

Definition

The real number $ad - bc$ is the *determinant* of matrix $M = \begin{bmatrix} a & b \\ c & d \end{bmatrix}$. (We can allow the entries to be complex or require that they be rational.)

The square matrix $I = \begin{bmatrix} 1 & 0 \\ 0 & 1 \end{bmatrix}$ is the *identity matrix* since for any 2-by-2 matrix M, $MI = IM = M$.

(continued)

When the determinant $ad - bc$ of M is not zero, then the matrix $\dfrac{1}{ad - bc}\begin{bmatrix} d & -b \\ -c & a \end{bmatrix}$ is M^{-1}, the *inverse* of matrix M, in that $MM^{-1} = M^{-1}M = I$. Those square matrices which have inverse matrices are called *invertible*.

The *General Linear Group of Degree Two* is the set of invertible 2-by-2 matrices under the operation of matrix multiplication. It forms a group since, as we shall see, it satisfies the four properties of a group.

Theorem C.1 *Function composition is an* associative *operation. If follows that matrix multiplication in the* General Linear Group *is associative.*

Proof Let f, g, and h be functions whose range and domain are such that the composition $f \circ (g \circ h)$ is defined. Now, $f \circ (g \circ h)(x) = f \circ g(h(x)) = f(g(h(x)))$ and $((f \circ g) \circ h)(x) = (f \circ g)(h(x)) = f(g(h(x)))$, the same.

Let M, N, and P be 2-by-2 matrices. Let x denote a column vector of 2 entries, and let Mx be $f(x)$, Nx be $g(x)$, and Px be $h(x)$. So $g(h(x))$ is $N(h(x)) = NPx$, and $f(g(h(x)))$ is $MNPx$. Since $(f \circ g) \circ h = f \circ (g \circ h)$ then $(MN)P = M(NP)$. $\qquad\square$

Theorem C.2 *The General Linear Group is* closed *under matrix multiplication.*

Proof Recall that the matrices of the General Linear Group are those with inverse matrices. We need to show that if square matrices M and N have inverses, then so does matrix MN. We can verify that matrix $N^{-1}M^{-1}$ is the inverse of matrix MN: $(MN)(N^{-1}M^{-1})$ is, by associativity, $M(NN^{-1})M^{-1} = MIM^{-1} = I$. $\qquad\square$

We conclude that all the group properties hold for matrices of the General Group under matrix multiplication.

Proof, Simplified, of Poncelet's *Fourth Principle* D

Theorem D.1 *Any circle and line in the plane of the circle can be projected to another plane so the circle is projected to a circle and the projection of the line "passes to infinity." [84] (Assume the line does not meet the circle.)*

Proof See Fig. D.1. Draw the perpendicular from the given line to the center of the circle, meeting the circle in points A and B and the line in point M. We now look at the plane on M, A and B that is perpendicular to the given line. In that new plane, draw the circle with diameter BM. (We'll assume $AB > AM$.) On A, we construct a perpendicular to BM meeting the circle with diameter BM at V and Q. On that same circle mark W between B and V so $\angle VQW \cong \angle VBM$. This means that $WQ \parallel VM$.

We draw two planes perpendicular to plane MBV, one on line MV and the other, parallel to this one, on QW.

The cone whose base is the given circle, with diameter AB, and whose vertex is V, is sliced by the plane on QW in a circle, the subcontrary circle to the given circle, since $\angle VBA \cong \angle VQW$. Projecting from V, the original circle with diameter AB is projected to this subcontrary circle on the new plane on WQ. And the given line on M is projected to infinity since $VM \parallel WQ$. □

Fig. D.1 Simplified from Poncelet's proof, 1813, of the *Quatrième principe, Fourth Principle*. Figure pictured is perpendicular to the plane of the given circle, where AB is a diameter of the given circle

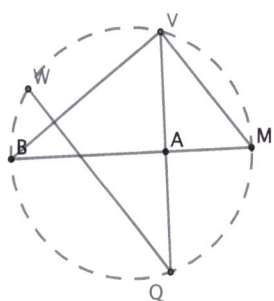

D Proof: Simplified of Poncelet's Fourth Principle

Alternative Solution, Due to Poncelet, 1822, of the Problem of Apollonius

E

Poncelet's *Traité*, Art. 269, *Fig.* 39, describes the construction of two circles which are each tangent, at the same time, to three given circles with centers C, C', C''. We assume the existence of such tangent circles, tangent, respectively, to the three given circle at A, A', A'', and at T, T', T''; those desired circles will be referred to as circles a and t. See Fig. E.1.

By Monge's Theorem: let the common external tangents to circles C and C', C and C'', and C' and C'' meet, respectively, at S'', S', and S. Then S'', S', and S are collinear. Call this line s.

Also from the chapter *Dilations*, when another circle is tangent to both circle a and circle t, the points of tangency are collinear with the center of the dilation pairing circles a and t. Therefore, lines AT, $A'T'$ and $A''T''$ are concurrent at the center of dilation, which we denote as R.

(*i*) R is the *radical center*, where the three common secants of the pairs of circles C, C', C'' meet. Why? From the *center*, X, of the dilation mapping circle a to circle t, each of the three circles C, C', C'' is tangent to both circle a and circle t. So by Theorem 17, Poncelet's Proposition 3 and Scholie of *Cahier* 1 of 1813, $XA \cdot XT = XA' \cdot XT' = XA'' \cdot XT''$. This means that the *power* of X with respect to the three circles is equal, so $X = R$ is the radical center of those three circles.

One consequence is that tangent lines to circle C at A and at T correspond as inverse homologues between circle a and t. Therefore, those two tangents meet on the common secant of circles a and t. This applies for circles C' and C'' as well. Therefore, the pole, P, of line s with respect to circle C lies on line AT. So A and T lie on line PR.

(*ii*) Now think about circles C and C', related by a dilation on center S''. This time, the third circle tangent to both circles C and C' is circle a. Its points of tangency with circles C and C' are A and A'. So line AA' lies on S''. Likewise, line TT' lies on S''. Lines AA' and TT' correspond as inverse homologues

C. Baltus, *Geometry by Its Transformations*, Compact Textbooks in Mathematics, https://doi.org/10.1007/978-3-031-72281-3

Fig. E.1 Poncelet's *Traité*, 1822, based on *Fig.* 39

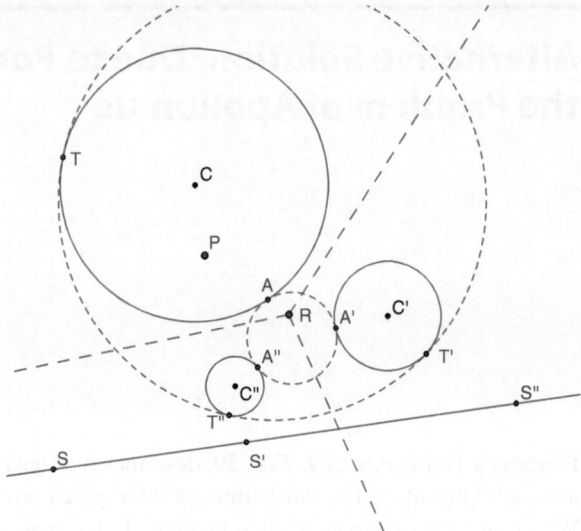

between circles *a* and *t*. So point *S″* lies on the common secant of *a* and *t*. The same thing holds for *S′* and *S*, so *s* is the common secant of circles *a* and *t*. And

(*iii*) by this, the pole of line *s* with respect to circle *C* must lie on line *AT*.

So we construct *P*, the pole of line *s* with respect to circle *C*, and join this pole to *R*. This gives us points *A* and *T*. In a similar fashion, we get points *A′* and *T′*, *A″* and *T″*, letting us construct circles *a* and *t*. □

The 1855 Argument of Moebius for Invariance of the Cross-ratio

In 1855, A. F. Moebius defined the *Kreisverwandtschaft*, **circle relationship**, as a one-to-one onto mapping of the extended complex plane that maps all circles and lines to circles and lines.

Theorem F.1 *Under a* circle relationship *relating the extended complex plane to itself, any four points A, B, C, D, are mapped to, respectively, points A', B', C', D', so that the cross-ratio CR(AC, BD) = CR(A'C', B'D').*

Proof The mapping is to pair one such plane, p, to another plane, p', where, by definition, any four points of a circle in one plane are paired with four points of a circle in the other. A line is a "circle" which includes the point at infinity. Moebius declared that one point M of the complex plane p was to be paired with the point at infinity, M', *unendliche entfernung*, in plane p'. And the point of infinity of plane p is to be called N, paired with point N' of plane p'. Further, he said his relationship was "continuous" in the sense that any two indefinitely near points, *unendlich naher Punckten*, correspond to indefinitely near points, and [Art. 7] that corresponding figures "due to their *unendlichen Kleinheit* are similar to each other." (If $M = M' = \infty$ then the relationship is a similarity.)

The heart of the argument is in Article 9, illustrated in our Figs. F.1 and F.2. Figure F.1 is from Moebius's 1855 article, and it is redrawn below as Fig. F.2. On the left is a triangle in plane p, where Q and P are indefinitely close to A, on sides AB and AM, respectively. Recall that M, in plane p, is mapped to infinity in plane p'. N, the point at infinity of p, lies on all lines, including the three sides of $\triangle ABM$. Since points A, Q, B, and N are collinear, then under the *Kreisverwandtschaft* their images lie on a circle, as on the right in Moebius's figure, keeping the same order. Since points A, P, M, and N lie, in order, on a line mapped to a line, then A' is between N' and P'. Since Q' is indefinitely close to A', then line $Q'A'$ is tangent at A' to circle $A'N'B'$. The vertical angle to $\angle Q'A'P'$ intercepts the arc $A'N'$, as

C. Baltus, *Geometry by Its Transformations*, Compact Textbooks in Mathematics, https://doi.org/10.1007/978-3-031-72281-3

Fig. F.1 Moebius Figure of
Article 9, [69], 1855

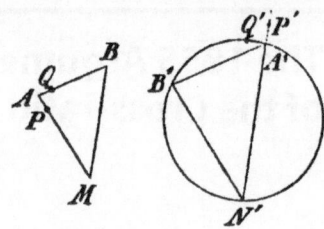

Fig. F.2 Moebius, based on
Figure of Article 9, [69],
1855

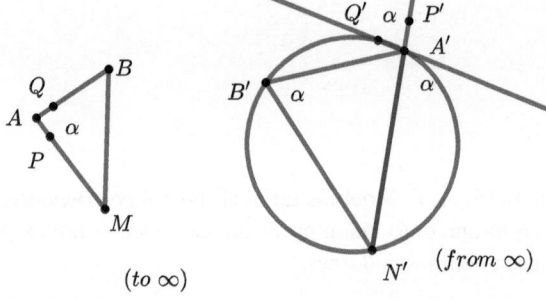

$(to \ \infty)$ $N' \quad (from \ \infty)$

does $\angle N'B'A'$, so those angles are congruent. Since indefinitely small angles are
preserved, then $\angle BAM \cong \angle QAP \cong \angle Q'A'P' \cong \angle A'B'N'$.

Then considering in the same way two points indefinitely close to vertex B, we
have $\angle ABM \cong \angle B'A'N'$.

Therefore, triangles ABM and $B'A'N'$ are similar, with that correspondence of
vertices. It follows, in Article 10, that $\dfrac{AB}{B'A'} = \dfrac{MA}{N'B'}$.

Let us consider two other points, C and D of plane p, and triangles CDM,
ADM, and CBM, and their images in plane p'. Treating points C, C', D, D' as
we treated points A, A', B, B' to get $\dfrac{CD}{D'C'} = \dfrac{MC}{N'D'}$, and then $\dfrac{AD}{D'A'} = \dfrac{MA}{N'D'}$ and
$\dfrac{CB}{B'C'} = \dfrac{MC}{N'B'}$. By multiplication, we get

$$\frac{AB}{B'A'} \cdot \frac{CD}{D'C'} \cdot \frac{D'A'}{AD} \cdot \frac{B'C'}{CB} = \frac{MA}{N'B'} \cdot \frac{MC}{N'D'} \cdot \frac{N'D'}{MA} \cdot \frac{N'B'}{MC} = 1.$$

Therefore,

$$\frac{AB \cdot CD}{AD \cdot CB} = \frac{B'A' \cdot D'C'}{D'A' \cdot B'C'} = \frac{A'B' \cdot C'D'}{A'D' \cdot C'B'}.$$

Thus, the cross-ratio of lengths is preserved.   □

Solutions to Selected Exercises

G

Chapter 1. Greek Background

3. This straightedge/compass construction follows directly from the Property (*ii*) of Exercise 1. Draw a circle with center A and radius length BC, and another circle with center C and radius AB. One of the points at which those circles meet will serve as D, since opposite sides of quadrilateral $ABCD$ are congruent.

9. Let the three circles be c_1, c_2, and c_3. Let circles c_1 and c_2 meet in points A and B, while circles c_2 and c_3 meet in C and D, and circles c_1 and c_3 meet in E and F. Let lines AB and CD meet in X. Because X is on AB, it has equal powers with respect to circles c_1 and c_2. As X is on CD, it has equal powers with respect to circles c_2 and c_3. Thus X has equal powers with respect to circles c_1 and c_3. So X also lies on EF.

    If lines AB and CD are parallel, it follows that the centers of the three circles are collinear. Hint: If two circles meet in points A and B, then the perpendicular bisector of AB lies on the centers of both circles. □

12. (Poncelet's solution.) Let lines AB and m meet at P. Draw a circle on A and B, and then construct a tangent from P to the circle, whose length we call s. Plot point S on m so $PS = s$. The circle we desire can be the one on S, tangent to line m, and on A. Why is B on this new circle? $PB \cdot PA = s^2$, and this equation holds only for the given B on line PA.   □

14. See Fig. G.1.

© The Author(s), under exclusive license to Springer Nature Switzerland AG 2025
C. Baltus, *Geometry by Its Transformations*, Compact Textbooks in Mathematics,
https://doi.org/10.1007/978-3-031-72281-3

Fig. G.1 Solution to our
Exercise 14, Problem 98,
from Hadamard 1906. It uses:
Inscribed Angle Thm, Alt.
Int. Angles of parallel lines,
tangents to a circle from an
outside pt. are $\cong$. Angles
marked α are congruent, etc.

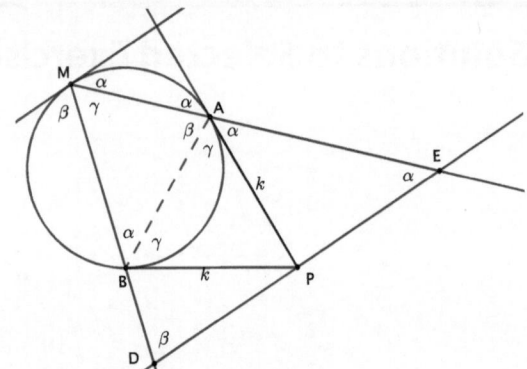

16. When $r < AB$, mark Q on line AB so $AQ = r$. Then on Q construct a parallel
    to AX. The point at which the new line on Q meets line BX can be Z. Then on
    Z construct a parallel to AB. Then A, Q, Z are three vertices of a parallelogram.
17. Take $\triangle XYW$ cut by line KBD and apply Menelaus's Theorem, then take
    $\triangle ZYW$ cut by line KCA and, again, apply Menelaus's Theorem. The desired
    equation follows from these two equations.
18. Suppose triangle ABC has Cevians AX, BY, CZ, either two or none outside
    $\triangle ABC$, where $AZ \cdot BX \cdot CY = AY \cdot BZ \cdot CX$. Then the Cevians are concurrent.

Proof Let Cevians AX and BY meet at S. Draw line CS meeting opposite side AB
in Z^*. By Ceva's Theorem, $AZ^* \cdot BX \cdot CY = AY \cdot BZ^* \cdot CX$. With the previous
equation, we have $\dfrac{AZ^*}{BZ^*} = \dfrac{AZ}{BZ}$. This is only possible if Z and Z^* are the same point
on line AB.

21. See Fig. G.2 of this appendix. $m\angle ADB = m\angle ACB = (m\ \overset{\frown}{AB})/2$ by the
    Inscribed Angle Theorem. By Euclid's Exterior Angle Theorem, $m\angle AFB >$
    $m\angle ADB$ and $m\angle ADB > m\angle AGB$.

$\square$

Fig. G.2 Circle of points D
such that $m \angle ADB =$
constant k or $\pi - k$

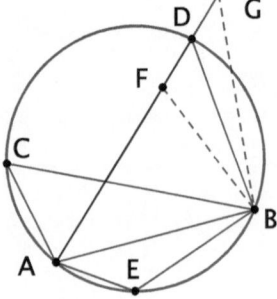

23. Let the center of the circle be C. Let the sliding angle have vertex X, while tangents from X meet the circle in A and B. Then $\triangle ACX \cong \triangle BCX$ by SSS. Angles CBX and CAX are both right angles. $m\angle AXB$ is independent of the choice of X, so $m\angle XCA = m\angle XCB$ is also independent of the choice of X. Thus, independent of the choice of X, all triangles ACX are congruent to each other, so length CX is the same. So X lies on a circle with center C and radius CX. $\square$

24. Let the quadrilateral be $ABCD$. Since it is inscribed in a circle, then $m\angle A + m\angle C = \pi$. Let tangents from A intercept minor arc a of the inscribed circle, etc. By Theorem 10 (iii), $m\angle A + m\angle C = \pi = (b+c+d-a+b+a+d-c)/2$. This means $b + d = \pi$. By Theorem 10 (i), the chords cutting opposite arcs b and d meet in a right angle.

Chapter 2. Dilations

1. (i) The dilation maps a line to a parallel line, so the image has slope 3. Point $(0, -2))$ is mapped to $(-6, 1)$. Answer: $y = 3x + 19$.

   (ii) The circle has center $(2, 0)$ and radius 2. The image has center $(2, 9)$ and radius 8, so the equation is $(x - 2)^2 + (y - 9)^2 = 64$.

2. We need parallel radii. For circle c_1, we can use radii to point $A = (0, 2)$ and to $B = (0, -2)$, and for circle c_2, the radius to point $D = (10, 7)$. One center of dilation will be the point where line AD meets the $x - axis$. AD has equation $y - 2 = 0.5x$; it meets the $x - axis$ at $E = (-4, 0)$. The other center of dilation is at the meeting point of line BD and the $x - axis$. BD has equation $y + 2 = 0.9x$; it meets the $x - axis$ at $(20/9, 0)$.

   To find the tangent to circle c_1 from $E = (-4, 0)$, which will be the common tangent, we first find the equation of the circle, c_3, whose diameter joins E to the origin:

   $(x + 2)^2 + y^2 = 4$. The points of tangency are the meeting points of circles c_1 and c_3. By the two equations, $4x + 4 = 0$, so $x = -1$. Then $y = \pm\sqrt{3}$. So one common tangent line is $y = \dfrac{\sqrt{3}}{3}(x + 1)$.

3. The quadrilateral whose sides are lines m, n, m' and n' is a parallelogram. The diagonals of a parallelogram bisect each other. Let the midpoint of diagonal XY be called Z. The dilation with center Z and scale factor -1 interchanges X and Y. Since a dilation maps a line to a parallel line, m is mapped to m' and n is mapped to n'.

4. For the first circle, construct the bisector of $\angle AXB$, then drop a perpendicular from a point of that bisector, C, to meet a side of the angle in a point M. The first circle can be with center C and lying on M. Draw line PX, and let Q be one of the points at which XP meets circle C. Apply a dilation with center X that maps Q to P. (There will be two solutions.)

6. Let A, B, and C be the three points, which we assume are not collinear. Let the perpendicular bisectors of sides AB and BC meet at point L. By an Exercise 8 of the previous chapter, the distances LA, LB, and LC are equal. This means

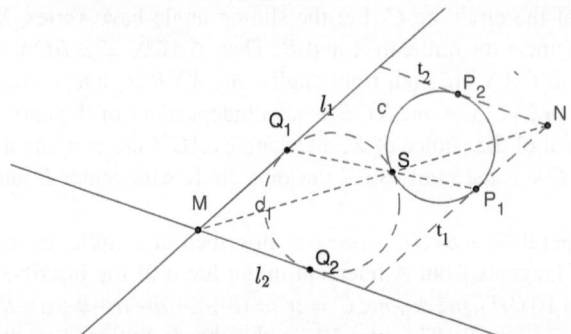

Fig. G.3 Exercise 7. Yaglom's solution to the Line-Line-Circle construction problem

L also lies on the perpendicular bisector of side AC. Since $LA = LB = LC$, then the circle with center L and lying on A is also on B and C. $\square$

7. (Yaglom, [106, p. 99]) See Fig. G.3. Let the given lines meet at M. Construct a line t_1 parallel to l_1, tangent to c (at P_1), and construct a line t_2 parallel to l_2, tangent to c (at P_2). Let the two new parallels meet at N. Let line MN meet c at S. Let line P_1S meet line l_1 at Q_1.

Let ϕ be the dilation whose center is S and which maps P_1 to Q_1. Since a dilation maps a line to a parallel line, line t_1 is mapped to line l_1. And so N is mapped to M and, then, line t_2 is mapped to l_2. Dilations map tangents to tangents and map circles to circles, so circle c is mapped to the circle, c_1, on Q_1 and Q_2 and tangent at those points to l_1 and l_2. The tangent line to circle c at S is mapped to itself, tangent at S to circle c_1. So c_1 is the circle we seek.

Chapter 4. Transformations as Listed by Moebius

2. In the equation of the circle, divide all three terms by a^2 and then replace y by $\dfrac{ay}{b}$. The resulting ellipse has the area of the circle multiplied by $\dfrac{b}{a}$. (The new $y - value$ is $\dfrac{b}{a}$ times the original y.) Therefore the ellipse has area $\pi a \cdot b$.

3. Let l and m be parallel lines. Suppose their images, l' and m', meet at a point A'. Since a transformation is one-to-one, there is exactly one point A mapped to A', and just one line l and one line m mapped to l' and m', respectively. A must lie on l and on m, so lines l and m meet. We have a contradiction.

Chapter 5. Background for Homology

1. Subtraction of the equations gives $2x + 6y + 9 = 0$, or $y = -\dfrac{x}{3} - \dfrac{3}{2}$. For an alternate method, both circles are tangent to the $y - axis$, at $y = 0$ and $y = -3$. So there are equal tangents to the circles from $(0, -\dfrac{3}{2})$. That point is on

the common secant. The common secant is perpendicular to the line joining the centers of the circles, $(2, 0)$ and $(1, -3)$, so the common secant has slope $-\frac{1}{3}$.

2. Following Suggestion 1, S, A_1, and A_2 will be collinear. For Suggestion 2, the three circles are the given circles and the circle sought. One of the common secants is the tangent to c_1 at A_1. The point at which this common secant meets the common secant of the two given circles will lie on the tangent at A_2.

4. Points A', B', C', D' have $x - coordinates$ $8, 2, -7, -16$, respectively. The cross-ratios are both $\frac{5}{4}$.

Chapter 6. Plane-to-Plane Projection

1. *Solution by Poinsot.* We consider a horizontal plane, such as a tile floor, and a vertical plane, called the *tableau*, which we can think of as that of a painting on an easel, perpendicular to the floor. The given lines AB and CD and point E are drawn on the tableau. Draw two lines on the tableau, one meeting AB at f and CD at h, and a second line meeting AB at g and CD at k, so fk and hg meet at the given point E. Let the two lines meet at point I on the tableau, on neither AB nor CD.

Draw a third line on I, meeting AB at m and CD at n, and let $gn \cap km$ be point T. Then line ET will be concurrent with AB and CD. Our Fig. G.4, based on Poinsot's solution, displays the figure on the tableau.

What does this represent on the floor? Regard the line on I and on the point at which AB and CD meet to be the vanishing line. The quadrilaterals $fgkh$, $gmnk$ and $fmnh$ on the floor are all parallelograms, since the projections of opposite sides meet the tableau on the horizon line. Since the diagonals of a parallelogram bisect each other, it follows that the line on the floor corresponding to line ET is parallel to the lines corresponding to AB and CD. Thus, AB, CD, and ET are concurrent on the extended tableau. (The reader familiar with Desargues' Theorem will recognize a shorter proof: The corresponding sides of $\triangle fEh$ and $\triangle mTn$ meet in collinear points so the lines joining corresponding vertices are concurrent.) $\square$

Fig. G.4 Figure to accompany the solution, May 1807, by Loius Poinsot, in *Correspondance sur l'École Imperiale Polytechnique*

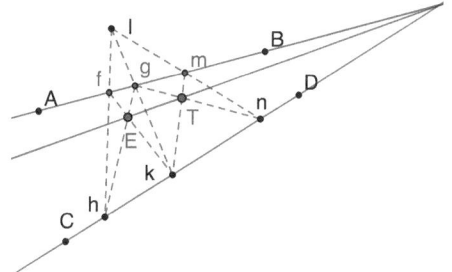

Fig. G.5 Solution
Exercise 4. Desargues'
Theorem Dual, after a
projection so
$AB \parallel A'B'$, $AC \parallel$
$A'C'$, $BC \parallel B'C'$. Apply
Poncelet's General Lemma,
1813

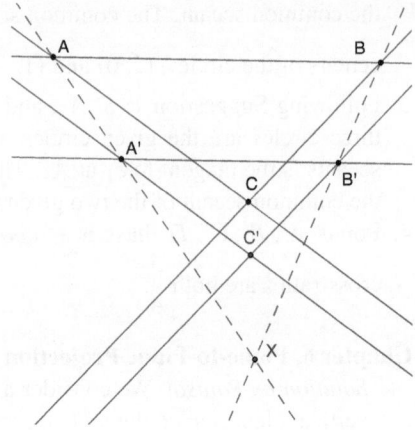

2. Given: (*i*) $BL \parallel DN$, $BM \parallel DP$, and
   (*ii*) lines LN, BD, MP meet at H.
   Show: $\triangle BLM \sim \triangle DNP$ and $LM \parallel NP$.
   Due to the parallel sides, with the Side Splitter Theorem applied to triangles
   LBH and MBH, we have $LH : NH = BH : DH = MH : PH$. Therefore,
   $LM \parallel NP$. We can then show that $\angle LBM \cong \angle NDP$ and $\angle BLM \cong \angle DNP$.
   It follows that $\triangle MBL \sim \triangle PDN$, by AA.

4. See Fig. G.5. By projecting line LMN to the line at infinity, we have, keeping
   the names unchanged under the projection, $AB \parallel A'B'$, $AC \parallel A'C'$, $BC \parallel$
   $B'C'$. Let AA' meet BB' at X. The corresponding sides of triangles ABC and
   $A'B'C'$ are parallel, so by Poncelet's General Lemma which followed *Cahier* 1
   Prop. 1 of 1813, the lines on corresponding vertices are concurrent. Concurrent
   lines are mapped to concurrent lines under projection, so in the original figure
   lines AA', BB', CC' are concurrent.

6. *Solution* of (*ii*) Consider hexagon $ADBECF$ inscribed in the given conic
   section. Opposite sides AD and EC meet at a point T, opposite sides DB and
   CF meet at a point U, and opposite sides BE and AF meet at a point V. By
   Pascal's Theorem, T, U, and V are collinear. Let D, E, and F move along the
   conic section approaching A, B, and C, respectively. In the limit, we have the
   tangent at A, B and C meeting their respective opposite sides of the triangle at
   collinear points.

   (*iii*) Given quadrilateral $ABCD$ with an inscribed conic tangent to sides
   AB, BC, CD, DA at, respectively, X, Y, Z, W. Then diagonals AC and BD
   meet at $XZ \cap YW$. Suggestion for the illustration: Let the conic be a circle.

7. See Fig. G.6, based on Poncelet's *Fig.* 33 of 1822. Poncelet used a circle for
   the conic section. Assume no three of A, B, C, D, E are collinear. (Otherwise,
   the "conic" would be one or two lines.) Let AB meet ED at L, and let BC meet
   line u at K. Draw the line LK and let line CD meet line LK at H. Then the
   point we seek is $F = u \cap AH$.

Fig. G.6 Solution of
Exercise 7, based on
Poncelet's *Fig.* 33, 1822

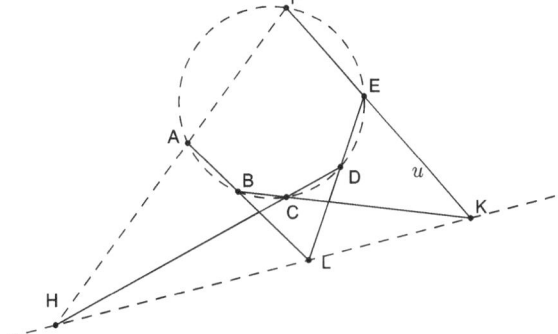

Note that as line u turns on E, we trace out the entire conic section. This is an argument that five points determine a conic.

9. Partial solution. Line a, the tangent at A, is the polar of A. Let a line, z, not a, lie on A. Let z meet the conic again at X. The tangent at X meets a at Z, the pole of z.

10. See Construction L, in Chap. 5.

Chapter 7. Homology

3. Any line on S meets the axis in a point K. Since K and S are fixed, that line is mapped to itself. For the converse, take a point K on the fixed line, l. The line SK is mapped to itself and line l is mapped to itself, so K, the intersection of the two fixed lines, must be fixed.

4. If there are two centers, S_1 and S_2, then any point X not on line S_1S_2 lies on two fixed lines, XS_1 and XS_2, so X' lies on those same lines, and the only such point is X. Now that we know that any point not on line S_1S_2 is fixed, we can use a similar method to show that any point X on line S_1S_2 is also fixed.

5. Solution of the first part. Line XX' meets the axis in a fixed point K. Since X and K are mapped to points on XX', the line XX' is mapped to itself.

6. (i) Since P is not fixed, line PP' is mapped to itself, and so, has a fixed point K on l. K cannot be a center so some line KQ is not fixed. As before, line QQ' is fixed and does not lie on K. Therefore point $PP' \cap QQ'$ is not on l but is fixed.

7. See Fig. G.7. (i) B' will lie on line SB. We take another line on B whose image we can construct: line AB. Let AB meet the axis at B_1. Since B_1 is fixed, the line on B_1, A, and B is mapped to line B_1A', on which B' must lie.
(ii) To find C on line SB, we need to first find C_1 on the axis so line $A'C_1$ is parallel to SB. Then C will be $C_1A \cap SB$.

8. See Fig. G.8.

9. See Fig. G.9.

11. Draw a line on S meeting line m in point M. M' will be $SM \cap m'$.

15. Solution of b.: ϕ_1 is a collineation with a center, S_1, so it maps the center, S_2, of ϕ_2 to the center, S_3, of $\phi_1 \circ \phi_2$. Under a homology, a point and its image are collinear with the center of the homology, so S_3 is collinear with S_1 and S_2.

Fig. G.7 Solution,
Exercise 7. Under homology
with center S, with given A
and A', find B' and find C on
line SB to be mapped to
infinity

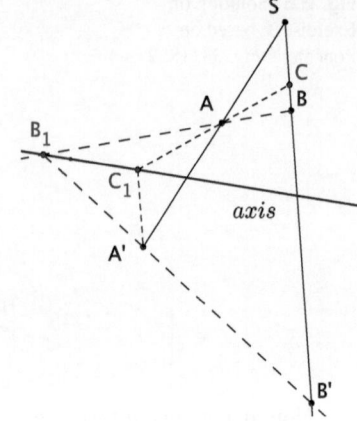

Fig. G.8 Solution.
Exercise 8. Homology with
axis CB mapping A to A'

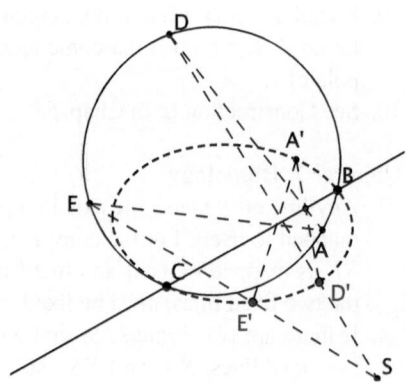

Fig. G.9 Solution.
Exercise 9. Homology with
axis CB mapping A to A'

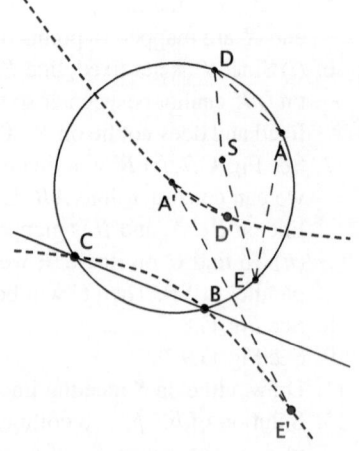

Fig. G.10 Based on Steiner's
Figure 38, 1832

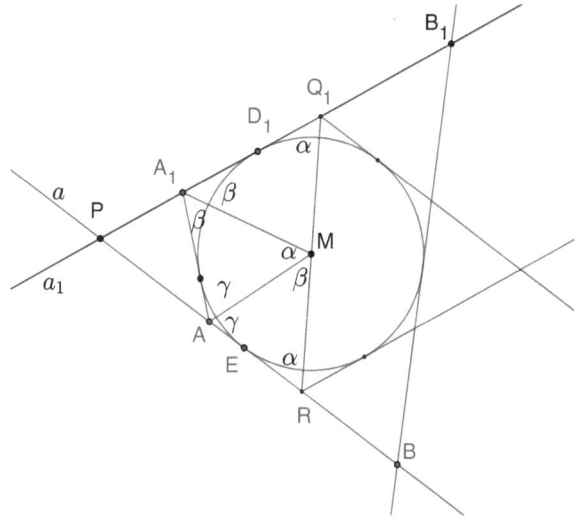

Chapter 8. Homogeneous Coordinates

1. $3x + 4y + 5z = 0$.
2. $(1, 3, 0)$.
3. The point on line $x + 3y + 2z = 0$ which lies at infinity is $(3, -1, 0)$. We need to find real numbers a, b, c so points $(2, 1, -2)$ and $(3, -1, 0)$ lie on line $ax + by + cz = 0$ in homogeneous coordinates. We find the answer: $2x + 6y + 5z = 0$.

Chapter 9. Projective Geometry

1. See Fig. G.10, the same as Fig. 9.14 of Chap. 9. For (*i*), the angles, marked β, at A_1 are equal since A_1M is a radius extended, and A_1Q_1 and A_1A are tangents. Likewise for the angles γ at A. As noted before, the measures of the four angles of quadrilateral Q_1A_1AR sum to 2π, from which it follows that $\alpha + \beta + \gamma = \pi$. So $m\angle Q_1A_1M = m\angle RMA = \beta$. The three triangles are similar by AA, and we get the angle measures in Fig. G.4. (*ii*) follows from the similar triangles; in particular, $AR \cdot A_1Q_1 = (MR)^2$.
   (*iii*) By corresponding reasoning, $BR \cdot B_1Q_1 = (MR)^2 = (MQ_1)^2$. Since, then, $AR \cdot A_1Q_1 = BR \cdot B_1Q_1$, it follows that $CR(AB, R \infty) = CR(A_1B_1, \infty Q_1)$.
   (*iv*) This means that this pairing of points of line a with points of a_1 is projective since the projective pairing defined by pairing A, R, ∞ with A_1, ∞, Q_1, respectively, would have to pair B with B_1.
3. The pencils at O and O' are projectively related so the set of points of intersection of corresponding lines forms a conic section. Why are they projectively related? The angle between consecutive marked lines in each pencil is half the intercepted arc, which is constant. The cross-ratios of corresponding sets of lines depend on equal angles, so are equal.

4. Take a line on point U that meets line u in A and line u_1 in A_1. This pairing of A and A_1 projectively relates lines u and u_1, and this relation projectively relates the pencils on S and S_1 by pairing line SA with line S_1A_1. In this last pairing, line SS_1 is paired with itself. By the La Hire-Steiner Theorem, the pencils on S and S_1 are in perspective, which means that corresponding lines meet in collinear points.

5. One solution. We take a harmonic set on line AB, say the pair A and B, and harmonic conjugates with respect to A and B, namely the midpoint, M, of segment AB and infinity on line AB. Take S on line CM so SD is parallel to AB. The lines SA, SB, SC, SD meet line AB in a harmonic set, so those four lines are a harmonic set.

6. Let the five points be A, B, C, D, E, with no three collinear. Let A and B be the two centers of two projectively related pencils: pair lines AC, AD, AE with lines BC, BD, DE, respectively. This completely determines a projective pairing of the lines on A with the lines on B, for any other line on A can meet that conic only at the one point where it meets the corresponding line on B. Thus, there is a conic section on those five points.

   On the other hand, take a conic section which includes no straight line, and take five points A, B, C, D, E lying on that conic. By the procedure used above, for any other line u on A (except the tangent) there is exactly one other point F of the conic on that line, and BF is the line of the pencil on B which corresponds to u.

7. See Fig. G.11. C' is $tan A \cap tan B$. A' is $tan B \cap tan C$, i.e., the point at infinity of $tan B$, and B' is $tan A \cap tan C$, i.e., the point at infinity of $tan A$. Line AA' is on point A and parallel to the axis of the parabola, since the axis lies on the point at infinity of the parabola. Line BB' is the line on B parallel to $tan A$. They meet at a point P. $APBC'$ is a parallelogram whose diagonals meet at M. This theorem follows: Given points A and B on a parabola (true for any conic), the midpoint of segment AB and the point at which the tangents at A and B meet lie on a diameter of the parabola. (In a parabola, the diameters are the lines parallel to the axis of the parabola.)

Fig. G.11 Problem 7
Solution, Parabola

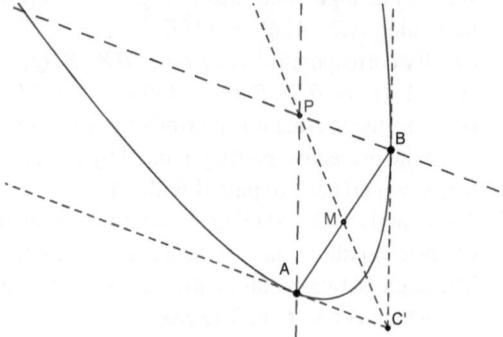

8. Lines AB and $A'B'$ will be interchanged by α, as will lines AB' and $A'B$. So each pair meets on a fixed point at infinity. Further, lines AA' and BB' are mapped to themselves, so they meet at a fixed point, M, the midpoint of segment AA'. So we can define a homology ϕ by setting the center to be M, the axis as the line at infinity, and mapping A to A'. Now show that B and B' are exchanged by ϕ and that A' is mapped to A. Because parallel lines will be mapped to parallel lines, lines AB and AB' will be mapped to lines $A'B'$ and $A'B$, respectively, and B will be mapped to a point on line BM, so B is mapped to B'. B' is mapped to a point on line $B'M$ and since line AB' is mapped to line $A'B$ then B' is $B'M \cap A'B$, which is point B. In a similar way, we see that A' is mapped to A. Since ϕ agrees at points A, B, B', and A' with collineation α, then ϕ is that same collineation.

Chapter 11. Inversion

2. The figure is mapped to itself, although not fixed pointwise.
5. If the center of inversion is O and the first inversion has radius of inversion r_1 and the second inversion has radius r_2, and point A is mapped to A' by the first inversion, and A is mapped to A'' by the second inversion, then $\dfrac{OA''}{OA'} = \dfrac{r_2^2}{r_1^2}$. Since this ratio holds for all points A, the images are related by a dilation, with scale factor $\dfrac{r_2^2}{r_1^2}$.
6. So that the circle is mapped to a line, the center of inversion must lie on the circle. First case: The line is a secant, cutting the circle at points A and B. Let CD be the diameter perpendicular to AB. We can let the circle of inversion have center C and radius CA, or we can let circle of inversion have center D and radius DA. Second case: The line does not meet the circle. Let C be the nearest point of the circle to the line and D the farthest point of the circle from the line, and M the point of the given line on line CD. Let the circle of inversion have center D and radius $\sqrt{DC \cdot DM}$.
7. See Fig. G.12. Suppose point A_2 has been found. Consider an inversion with center at O, one of the points of intersection of the two given circles. The two circles are mapped to lines which meet at the image, X, of point O_2, the other point at which the given circles meet. One of those two lines is on XA'_1 and the other is on XA'_2. Place A'_2 on this second line so $XA'_1 = XA'_2$, so there is a circle on A'_1 and A'_2 tangent to the two lines. We conclude that A_2 is the image of A'_2 by the inversion. (Note: In the Exercises at the end of the Dilations chapter, another methods of finding A_2 is suggested.)
8. See Fig. G.13. A dilation maps a circle to a circle and maps a tangent line to one circle to a tangent line to the second circle. m, a line on the center of the dilation, must be mapped to itself. Circle c_3, tangent to m at C and lying on B must be mapped to the unique circle which is tangent to m at C and lying on D, which is circle c_4.

For (*iii*), note that B is mapped to D. A dilation maps a line to a parallel line and a tangent to a tangent, so the tangents at B and D must be parallel.

Fig. G.12 Inversion,
Exercise 7

Fig. G.13 Circles c_3 and c_4
related by the dilation with
center C pairing points B and
D

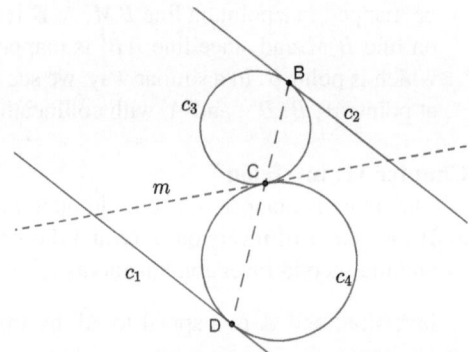

9. (from Yaglom) Apply an inversion with center A. Let B' denote the image of B,
   etc. The given circle on A and B and the given circle on A and D are mapped
   to lines which cannot meet since the two given circles meet only at A. The given
   circle on B and C will be mapped to a circle tangent at B' to the new line on B'
   and the circle on D and C will be mapped to a circle on D' which is tangent to
   the new line on D', and these two new circles will be tangent to each other at C'.
   By Exercise 8, B', C' and D' are collinear. A', at infinity, is on all lines so the
   original points A, B, C, and D lie on a circle.

Chapter 12. Moebius Transformation

1. (i) Let $w_1 = 1$ be mapped to 1, $w_2 = -1$ mapped to -1; and $w_3 = i$ mapped to
   ∞. The equation

$$\frac{(w_1 - w_2)(w_3 - z)}{(w_1 - z)(w_3 - w_2)} = \frac{(w_1' - w_2')(w_3' - z')}{(w_1' - z')(w_3' - w_2')},$$

is $\dfrac{(1 - -1)(i - z)}{(1 - z)(i + 1)} = \dfrac{(1 - -1)(\infty - z')}{(1 - z')(\infty + 1)}$.

The right side is $\dfrac{2}{1 - z'}$. The left is $\dfrac{2(i - z)}{(1 - z)(i + 1)}$.

Solving for z' gives

$$z' = \frac{1 - iz}{z - i}.$$

(ii) By the matrix form, with $ad - bc = -2$, and $f(z) = \begin{bmatrix} -i & 1 \\ 1 & -i \end{bmatrix} \begin{bmatrix} z \\ 1 \end{bmatrix}$,

$f^{-1}(z) = \dfrac{1}{2} \begin{bmatrix} i & 1 \\ 1 & i \end{bmatrix} \begin{bmatrix} z \\ 1 \end{bmatrix}$.

2. Solution of (i) Since the point at infinity is fixed, any line and its image meet at the point at infinity, meaning they are parallel. (ii) By part (i), the sides of $\triangle ABC$ are mapped to straight lines. Since $CR(AB, C\infty) = CR(A'B', C'\infty)$, then, due to ∞, $\dfrac{AC}{BC} = \dfrac{A'C'}{B'C'}$ and $\angle BCA \cong \angle B'C'A'$. So by the SAS Triangle Similarity Theorem, $\triangle ACB \cong \triangle A'C'B'$. $\square$

3. $A' = \dfrac{-130 + 150i}{485}$, $B' = -.5$, $M = -2$, $N' = 2$. $AM = 2.20227$, $AB = 0.223607$, $MB = 2$, $B'N' = 2.5$, $B'A' = 0.253837$ and $N'A' = 2.27038$. One can show that corresponding sides are (nearly) proportional to prove that the triangles are (nearly) similar.

4. $CR(AB, CD) = 7/6$ and $CR(AC, BD) = -1/6$. The four points are in order $A - B - C - D$ on a circle so they can be mapped by a Moebius transform to collinear points in the same order. In the case $CR(AB, CD)$ the cross-ratio is positive and in the case $CR(AC, BD)$ a negative cross-ratio results.

5. The formula used in the solution to Exercise 1 yields, for $w_3 = 1 + 2i$, $z' - w_3 = K(z - w_3)$ for a complex constant K. This means that the transformation rotates the plane by $arg(K)$ about the fixed point w_3 and then carries out the dilation with center w_3 and scale factor $|K|$. We can find K by replacing z by 2 and z' by 3.

6. (a) By the invariance of the cross-ratio, $CR(AZ, \infty w) = CR(\infty Z, B_1 w')$. So $\dfrac{(A\infty)(Zw)}{(Aw)(Z\infty)} = \dfrac{(\infty B_1)(Zw')}{(\infty w')(ZB_1)}$. Simplifying due to ∞, we get $\dfrac{(Zw)}{(Aw)} = \dfrac{(Zw')}{(ZB_1)}$.

Therefore, $w' - Z = (ZB_1)\dfrac{w - Z}{w - A}$.

(b) Replace w by S and w' by S'. This means $\dfrac{Z - S}{A - S} = \dfrac{Z - S'}{Z - B_1}$. Note that the quadrilateral $SAZB_1$ is a parallelogram. So vectors $A - S$ and $Z - B_1$ are equal, making our equation $Z - S = Z - S'$. Thus $S' = S$. $\square$

Chapter 15. Isometries and Dilations in French Schoolbooks

1. Solution of b. In the Fig. 71, mark X at the intersection of lines AB and A_1B_1, which we call, respectively, lines m and m_1. Let θ be $m\angle ASA_1$, which is also $m\angle BSB_1$. Rotation by θ rotates any line by θ. Therefore, angle AXA_1 is the supplement of angle θ. This means that quadrilateral $SAXA_1$ is cyclic, so there is a circle on those four vertices. As segments SA and SA_1 are congruent, they intercept congruent arcs of circle AXA_1, so inscribed angles AXS and A_1SX are congruent. In other words, SX is the bisector of the angle formed by lines m and m_1.

2. Proof of (v) of the Isometry Theorem. Note that the conditions of (v) determine the isometry since the image of any $\triangle ABZ$ for Z not on line AB is determined by isometry and orientation properties. Now, let the perpendicular bisectors of segments AA' and BB' meet in point S. This means $SA = SA'$ and $SB = SB'$. Also, $AB = A'B'$. So $\triangle ASB \cong A'SB'$, so $\angle ASB \cong \angle A'SB'$. Adding or subtracting the measure of $\angle BSA'$ gives $\angle ASA' \cong \angle BSB'$. Therefore, the given isometry is the rotation about S by $\angle ASA'$. So S is fixed.

3. Solution for SAS triangle similarity. Let ABC and DEF be the given triangles, with $\dfrac{DE}{AB} = \dfrac{EF}{BC}$, and $\angle B \cong \angle E$. Let $k = \dfrac{DE}{AB}$. Apply to $\triangle ABC$ the dilation with center B and scale factor k, where we call the image triangle $A'BC'$. Since a dilation multiplies segment length by k, then $A'B = AB \cdot k = DE$ and $C'B = FE$. A dilation fixes angle measure, so by the SAS Triangle Congruence Theorem, $\triangle A'BC' \cong \triangle DEF$. Since the image of $\triangle ABC$ (under a dilation) is congruent to $\triangle DEF$, then $\triangle ABC$ is similar to $\triangle DEF$. $\square$

5. $\begin{bmatrix} 0 & -1 \\ -1 & 0 \end{bmatrix}$.

6. $(-2, 4)/\sqrt{2}$, $(-5, 3)/\sqrt{2}$.

7. (a) The point at infinity in the direction perpendicular to line m. (b) The point at infinity in the direction of vector x.

8. Let θ be the angle with vertex S that rotates line m to line n. The rotation is by 2θ.

9. It is the dilation with center $(2, 3)$ and *ratio*, or *scale factor*, -2.

10. Suppose the isometry has no fixed point but it maps $\triangle XYZ$ to $\triangle X'Y'Z'$ reversing orientation. Let us translate so one vertex of the triangle, say X, goes to its image, X', giving us $\triangle X'Y^*Z^*$. Draw the bisector, m, of angle $Y^*X'Y'$ and reflect over m, sending side $X'Y^*$ to side $X'Y'$. Since orientation is reversed by the translation followed by a reflection, then Z^* is reflected to Z'.

Answers to Selected Exercises in Appendix 1. Additional Exercises

7. (Not from Yaglom.) (i) Let the tangents to the given circle at A and B meet at S. So $SA = SB$. Assume the circles sought have been drawn, tangent to each other at H. Applying the Common Secant Theorem to the three circles, a common tangent to the constructed circles lies on S. To see all the possible points H, note that any ray from S passing between A and B will work as the common tangent to the circles sought. The perpendicular at A to AS and the bisector of angle ASH will meet at the center of the circle tangent at A. Likewise for the circle tangent at B (Fig. G.14).

(ii) Let J be the midpoint of AH. Angle AJS is a right angle, so J ranges along a half-circle with diameter SA, from A until J lies on the bisector of angle ASB. If we apply a dilation with center A and scale factor 2, the half-circle with diameter AS becomes the half-circle on which H ranges, half the circle with center S that lies on A (and B). (Note that ray SH may be drawn outside the given circle.)

Fig. G.14 Exercise 7, from Yaglom, *Geometric Transformations* IV, p. 14

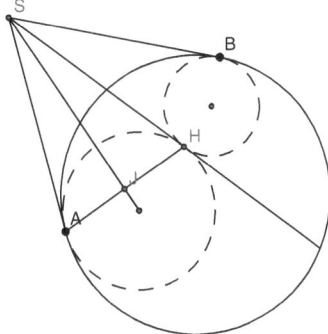

Yaglom provides two solutions that involve inversions. [107, p. 172]

8. The image of line LC is the line on L that is parallel to line SC. Let a line on S meet LC at B and meet the image of LC at B_1. Point $AB \cap A_1B_1$ will be fixed, so it lies on the axis.

9. The only dilation relating the two circles, with the given values of A, A', etc., has center S, where common tangents meet, and mapping A to B'. Since S is the center of inversion acting on circle c, the product $SA \cdot SA'$ is j^2 where j is the radius of the circle of inversion (not shown in the diagram). $SA \cdot SB = p$ is the power of S with respect to circle c. The dilation has scale factor k where $SB' = kSA$. With $SA \cdot SB = p$, then we find the scale factor of the dilation mapping circle c to circle c' to be $k = SB'/SA = SB' \cdot SB/p = j^2/p$.

10. (a) Corresponding sides must meet at collinear points J, K, and L. When sides AB and A_1B_1 meet at point J on the line at infinity, in order that J, K, and L be collinear, we need line KL to be parallel with AB and A_1B_1.

    (b) Consider triangles $B_1C_1S_1$ and $B_2C_2S_2$. Their corresponding sides meet in C, B, and, at infinity, where parallel lines B_1C_1 and B_2C_2 meet. Therefore, by Desargues' Theorem, triangles $B_1C_1S_1$ and $B_2C_2S_2$ are in perspective, from point $B_1B_2 \cap C_1C_2$. Now look at triangles $S_1B_1C_1$ and $S_2B_2C_2$. Their corresponding sides meet at B, C, and the point at infinity where parallel lines B_1C_1 and B_2C_2 meet, so these triangles, too, are in perspective from point $B_1B_2 \cap C_1C_2$. This means that lines B_1B_2 and C_1C_2 meet on line S_1S_2. So S_3 lies on m. $\square$

    Note that a dilation is a homology whose axis is the line at infinity and whose center is finite. The last Exercise of the Homology chapter is an alternate argument proving the claim of this Exercise 10*b*.

11. See Fig. G.15. With the projectively related pencils on S and S_1, pairing the lines which meet on the conic, then points of lines DA and DC are projectively related with D corresponding to itself, M paired with C, K paired with T, and A paired with R. This means lines DC and DA are related by a perspectivity. The center of that perspectivity must be $S_2 = RA \cap CM$. As K and T correspond, then line TK also lies on S_2. Therefore, the opposite sides of hexagon SBS_1CDA meet in collinear points, namely, T, K, and S_2.

Fig. G.15 Proof of Pascal's
Theorem, Dowling Fig. 38

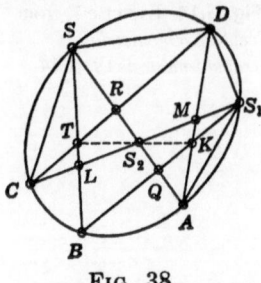

Fɪɢ. 38.

12. Take the circle on A, A', and B'. By the power of point I with respect to the new
    circle, $IB \cdot IB' = r^2 = IA \cdot IA'$ where r is the radius of inversion. Therefore,
    B is on the circle on A, A', and B'.
13. Part (i) $x^2 + y^2 - z^2 = 0$.   Part (iii) $(a, \pm b, 0)$.

Bibliography

1. Norbert A'Campo and Athanase Papadopoulos, On Klein's So-called Non-Euclidean geometry, [math.MG], arXiv:1406.7309v1, 2014.
2. Irving Adler, What shall we teach in high school geometry?, *Mathematics Teacher* 66 No. 3, 1968, 226–238.
3. A. V. Akopyan and A. A. Zaslavsky, *Geometry of Conics*, transl. A. Martsinkovsky, Providence, Rhode Island: American Mathematical Society, 2007.
4. Leon Battista Alberti, *De pictura*, 1435, in *On painting and and on sculpture. The Latin texts of De pictura and De statua*, ed. and transl. Cecil Grayson, London: Phaidon, 1972.
5. Apollonius of Perga, *Apollonii Pergaei conicorum libri quattuor . . .* , ed. and transl. Federicus Commandinus, Bononiae, 1566. Reissued with new diagrams: Pistorii, 1696.
6. Apollonius of Perga, *Conics*, transl. R. C. Taliaferro, Santa Fe, NM: Green Lion Press, 1998.
7. Archimedes, *The Works of Archimedes with the Method of Archimedes*, ed. T. L. Heath, Cambridge University press, 1912, reissued by New York: Dover, 2002.
8. Christopher Baltus, *Collineations and Conic Sections: An Introduction to Projective Geometry in its History*, New York: Springer, 2020.
9. Christopher Baltus, Poncelet's discovery of homology, *Historia Mathematica*, vol. 63, 2023, 1–20.
10. Giusto Bellavitis, Teoria delle figure inverse, e loro uso nella Geometria elementare, *Annali della Scienze del Regno Lombardo-Veneto* 6, 1836, 126–141.
11. Eugenio Beltrami, Saggio di interpretazione della geometria non-euclidea, *Giornale di Matematiche* 6, 1868, 285–315.
12. Johann Bernoulli, *Johannis Bernoulli, Opera Omnia* Vol. 4, Lausanne and Geneva: Bousquet, 1742.
13. Robert Bix, *Conics and Cubics: A Concrete Introduction to Algebraic Curves*, New York: Springer, 1998.
14. Rudolf Bkouche, Variations autour de la réforme de 1902/1905, in Hélène Gispert, *La France Mathématique*, La Société Mathématique de France (18670–1914), SFHST et SMF (1991), 181–214.
15. Émile Borel, *Géométrie: 1er et 2nd cycles*, Paris: A. Colin, 1910.
16. Charles-Julien Brianchon, Sur les surfaces courbes du second degré, *Journal de l'École Polytechnique*, Cahier 13, Tome 6, 1806, 297–311.
17. Charles-Julien Brianchon, Solution de plusieurs problèmes de géométrie, *Journal de l'École polytechnique*, Cahier 10, Tome 4, 1810, 1–15.
18. Charles-Julien Brianchon, Géométrie de la regle, *Correspondance sur l'École Impériale Polytechnique*, No. 5, 2e vol., 1813, 384–387.
19. Charles-Julien Brianchon, *Mémoire sur les Lignes du second Ordre*, Paris: Bachelier, 1817.
20. Constantin Carathéodorie, *Conformal Representation*, Cambridge University Press, 1932.
21. Lazare Carnot, *Géométrie de Position*, Paris: Duprat, An XI, 1803.
22. Lazare Carnot, *Essai sur la Théorie des Transversales*, Paris: Courcier, 1806.

23. Arthur Cayley, Sixth memoir on quantics, *Phil. Trans. Royal Soc. London*, Vol 149, 1859, 61–90.

24. Arthur Cayley, On the matrix $\begin{bmatrix} a & b \\ c & d \end{bmatrix}$ and in connection there with the function $\dfrac{ax+b}{cx+d}$, *Messenger of Mathematics* 9, 1880, 104–109.

25. Giovanni Ceva, De lineis rectis, Milan, 1678.

26. Herbert Oettel, Giovanni Ceva, in *New Dictionary of Scientific Biography*, Scribner's Sons, 2008.

27. Michel Chasles, *Traité de Géométrie Supérieure*, Paris: Bachelier, 1852.

28. Michel Chasles, *Traité des Sections Coniques*, Paris: Gauthiers-Villars, 1865.

29. Ganesh Prasad Chatterji, *Some Great Mathematicians of the Nineteenth Century*, Vol. 1, Benares, India, 1933.

30. H. S. M. Coxeter, *Introduction to Geometry*, New York: Wiley, 1969.

31. Arthur F. Coxford and Zalman Usiskin, *Geometry, a Transformation Approach*, Chicago: Laidlaw, 1971.

32. Arthur Coxford, Zalman Usiskin, Daniel Hirschhorn, *Geometry* in *The University of Chicago School Mathematics Project*, Glenview, IL: Scott, Foresman, 1991.

33. Richard Courant and Herbert Robbins, *What is Mathematics?*, New York: Oxford University Press, 1941.

34. Luigi Cremona, *Elements of Projective Geometry* 2nd edition, transl. by C. Leudesdorf, Oxford: Clarenden Press, 1893.

35. Girard Desargues, *Brouillon project d'une atteinte aux événements des rencontres d'un cône avec un plan*, in [44] (transl. Field) and original http://gallica.bnf.fr/ark:/12148/bpt6k105071b/f1.image Paris, 1639.

36. Girard Desargues, Note: - Extrait de la perspective de Bosse 1648, et faisant suite à la perspective de Desargues de 1636, in David Eugene Smith *A Source Book in Mathematics*, vol 2, New York: Dover, 1959, 307–309.

37. L. Wayland Dowling, *Projective Geometry*, New York: McGraw-Hill, 1917.

38. J. B. Durrande, Théorie élémentaire des contacts des circles, des sphères, des cylindres et des cônes, *Annales de Mathématiques Pure et Appliquées* Tome 11, 1820, 1–67.

39. Euclid, *Euclid's Elements*, edited and translated by David Joyce, https://mathcs.clarku.edu/~djoyce/java/elements/elements.html, 1998.

40. Leonhard Euler, *Introduction to analysis of the infinite, Book II*, 1748, transl. John Blanton, New York: Springer, 1990 (original 1748).

41. Leonhard Euler, De centro similitudinis [E693], *Leonhardi Euleri Opera Omnia* sub ausp. Soc. Scient, Nat. Helv. 1911- Series I vol. 26, 276–285 (original 1777).

42. Leonhard Euler, Solutio facilis problematum quorumdam geometricorum difficillimorum [E325], *Novi commentarii academiae scientiarum imperialis Petropolitanae* 11, (1765) 1767, pp. 12–14, 103–123. Reprinted in Opera omnia I. 26, pp. 139–157. Available online at EulerArchive.org.

43. L. A. S. Ferriot, *Annales des Mathématiques pures et appliquées*, Tome 2, 1811–1812, p. 180–182.

44. J. V. Field and J. J. Gray, *The Geometrical Work of Girard Desargues*, New York: Springer, 1987.

45. Henry George Forder, *Geometry*, London: Hutchinson University Library, 1960.

46. L. Gaultier, Mémoire sur les moyens généraux de construire graphiquement un circle déterminé par trois conditions et une sphère déterminée par quatre conditions, *Journal de l'École polytechnique* XVI, 1813, 124–214.

47. Joseph Diez Gergonne, Géométrie analitique. Théorie analitique des pôles des lignes et des surfaces du second ordre, *Annales de Mathématiques pures et appliquées*, tome 3 (1812–1813), 293–302

48. Joseph Diez Gergonne, Philosophie mathématique. Considérations philosophiques sur les élémens de la science de l'étendue, *Annales de Mathématiques pures et appliquées*, tome 16 (1825–1826), 209–231.

49. Joseph Diez Gergonne, Géométrie de situation, *Annales des Mathématiques*, vol 18, 1827–28, 149–154.
50. G. J. 'sGravesande, *Essai de perspective*, The Hague: veuve d'Abraham Troyel, 1711.
51. J. N. P. Hachette, Memoire sur le contact des spheres, in *Correspondance sur l'École Imperiale Polytechnique*, 1807 issue, in tome 1, 1808, 17–29.
52. Jacques Hadamard, *Leçons de Géométrie élémentaire*, 10th edition, Paris: Librairie Armand Colin, 1906.
53. Sir Thomas Heath, *Apollonius of Perga Treatise on Conic Sections*, Cambridge University Press, 1896; reissued by Palala Press, 2015.
54. Hermann von Helmholtz, Über die Thatsachen, welche der Geometrie zu Grunde liegen, in *Wissenschaftliche Abhandlungen*, Volume II, Leipzig: Barth, 1883, 618–639. Originally published in the Nachrichten von der Königl. Gesellschaft der Wissenschaften zu Göttingen, No. 9, 1868.
55. J. Henrici and P. Treutlein, *Lehrbuch der Elementar-Geometrie*, Leipzig: Teubner, 1881.
56. David Hilbert, *Grundlagen der Geometrie*, Leipzig: Tuebner, 1899.
57. Roger A. Johnson, *Modern Geometry; an elementary treatise on the geometry of the triangle and circle*, Boston: Houghton, Mifflin, 1929.
58. Felix Klein, Vergleichende Betrachtungen über neuere geometrische Forschungen, Erlanger: Deichert, 1872, translation by M. W. Haskell as A comparative review of recent researches in geometry, *Bull. New York Math. Soc.*, 2, 1892–1893, 215–240.
59. Felix Klein, Ueber die sogenannante Nicht-Euklidische Geometrie, *Mathematische Annalen* 4, 1871, 573- 625.
60. Philippe de La Hire, *Nouvelle Méthode en Géométrie pour les Sections des Superficies coniques et Cylindriques*, Paris 1673.
61. Philippe de La Hire, *Nouveaux Elements des Sections Coniques; Les Lieux Géométriques; La Construction, ou Effection des Équations*, Paris: André Pralard, 1679; Reissued: Whitefish, MT: Kessinger Publishing, 2009.
62. Philippe de La Hire, *Sectiones Conicae en novem libros distributae*, Paris 1685; French translation by Jean Peyroux, *Grand Livre des Sections Coniques*, Paris: Blanchard, 1995.
63. Adrien-Marie Legendre, *Élémens de Géométrie, avec des notes*. Paris: Firmin Didot, An II, 1794.
64. Colin Maclaurin, Letter from Mr. Colin Mac Laurin, in *Philosophical Transactions of the Royal Society*, Vol 39, Issue 439, 1735.
65. Charles Méray, *Nouveaux Éléments de Géométrie*, Paris: F. Savy, 1874.
66. France, Ministère de l'éducation nationale, *Bulletin administratif du ministère de l'instruction publique*, Volume 78, 1905.
67. August Ferdinand Moebius, *Der Barycentrische Calcul : ein neues Hülfsmittel zur analytischen Behandlung der Geometrie*, Leipzig: Barth, 1827.
68. August Ferdinand Moebius, Ueber eine neue Verwandtschaft zwischen ebenen Figuren, 1853, in [70] Vol. II, 205–218.
69. August Ferdinand Moebius Die Theorie der Kreisverwandtschaft in rein geometrischer Darstellung, 1855, in [70] Vol. II, 243–314.
70. August Ferdinand Moebius, *Gesammelte Werke*, Königlich sächsischen Gesellschaft der Wissenschaften, Leipzig: S. Hirzel, 1885–87.
71. Gaspard Monge, *Géométrie descriptive. Leçons donnée aux Écoles normals, l'an 3 de la République* (1797), Paris: Baudouin, an VII/1799.
72. Philippe Nabonnand, Les réformes de l'enseignement des mathématiques au début du XXe siècle, INRP; Vuibert, pp. 293–314, 2007, 978-2-7117-8981-8. hal-01083143.
73. Isaac Newton, *The Principia: Mathematical Principles of Natural Philosophy*, edited and translated by I. B. Cohen and Anne Whitman, based on the 1726 third edition, in Latin. Berkeley: University of California Press, 1999.
74. Shen Kangshen, John Crossley, Anthony W.-C. Lun, *The Nine Chapters on the Mathematical Art, Companion and Commentary*, Oxford University Press, 1999.

75. J. J. O'Connor and E. F. Robertson, biography of Jean-Victor Poncelet at https://mathshistory.st-andrews.ac.uk/Biographies/Poncelet/, 2008.

76. Pappus, *Pappi Alexandrini Mathematicae Collectiones*, trans. by Federico Commandino, Pisa: Apud Hieronymum Concordiam, 1588.

77. Blaise Pascal, *Essay pour les coniques*, 1640, https://www.persee.fr/doc/rhs_0048-7996_1955_num_8_1_3488 p 11–18, in English transl. in [44], 180–184.

78. Boyd Patterson, The origins of the geometric principle of inversion, *Isis*, Vol 19, No. 1, 1933, 154–180

79. Daniel Pedoe, *An Introduction to Projective Geometry*, New York: Pergamon Press, 1963.

80. Julius Plücker, Mémoire sur les contacts et sur les intersections des cercles, *Annales de Mathématiques pures et appliquées*, tome 18, 1827, 29–47.

81. Louis Poinsot, Problême de géométrie, solution reported by Hachette, *Correspondance sur l'École Imperiale Polytechnique*, May 1807, in tome 1, 1808, 305–306.

82. J. V. Poncelet, Problêmes de géométrie, *Correspondance sur l'École Imperiale Polytechnique*, 1811, in tome II, 1813, 271–274.

83. J. V. Poncelet, *Applications d'analyse et de géométrie qui ont servi, en 1822, de principal fondement au traité des propriétés projectives des figures, etc.*, 2 tomes, Paris: Mallet-Bachelier, 1862–64.

84. J. V. Poncelet, 7 cahiers de 1813–14, in [83], Tome 1, 1–441.

85. J. V. Poncelet, Essai sur les propriétés projectives des sections coniques, 1820, in [83], Tome 2, 365–441.

86. J. V. Poncelet, Théorèmes nouveaux sur les lignes du second ordre, *Annales de Mathématiques de Montpellier*, Vol VIII, 1817–1818, p 466, in [83].

87. J. V. Poncelet, *Traité des Propriétés Projectives des Figures*, Paris: Bachelier 1822.

88. J. V. Poncelet, Mémoire sur la théorie générale des polaires réciproques; pour faire suite au Mémoire sur les centres de moyennes harmoniques, *Journal für die reine und angewandte Mathematik*, 1829, 1–71.

89. Claudius Ptolemy, The Arabic version of Ptolemy's Planisphere or Flattening the Surface of the Sphere: Text, Translation, Commentary; ed. and transl. by J. L. Berggren and Nathan Sidoli, in *Sources and Commentaries in Exact Sciences* 8 (2007), 37–139.

90. Bernhard Riemann, Über die Hypothesen welche der Geometrie zu Grunde liegen, lecture of 1854 at the University of Göttingen, published *Abhandlungen der Königlichen Gesellschaft der Wissenschaften zu Göttingen*, vol. 13, 1868, 133–150.

91. George Salmon, *A Treatise of Conic Sections*, Third Edition, London: Longman, Brown, Green, and Longmans, 1855.

92. Simon Schama, *Citizens: A Chronicle of the French Revolution*, New York: Knopf, 1989.

93. François-Joseph Servois, *Solutions peu connues de différens problèmes de géométrie-pratique*, Bachelier: Paris 1804. Transl. and commentary by Salvatore Petrilli, *Servois's Solutions peu connues de différens problèmes de géométrie-pratique*, Boston: Docent, 2017.

94. Thomas Simpson, *Elements of Geometry; with their application . . .* , London: Nourse, 1760.

95. Robert Simson, *Apollonii Pergaei Locurum planorum, Libri II*, Glasgow, 1749.

96. Jacob Steiner, Einige geometrische Betrachtungen, *Journal für die reine und angewandt Mathematik*, Vol. 1, 1826, 161–184.

97. Jacob Steiner, Geometrische Lehrsätze, *Journal für die reine und angewandt Mathematik*, Vol. 2, 1827, 190–193.

98. Jacob Steiner, *Systematische Entwicklung der Abhängigkeit geometrischer Gestalten*, Erster Theil, Berlin: G. Fincke, 1832, in *Jacob's Steiner's Gesammelte Werke*, Erster Band, edited by K. Weierstrass, Berlin: G. Reimer, 1881, 229–460.

99. Lytton Straghey, *Eminent Victorians*, New York: Harcourt, Brace, 1918.

100. Glen Van Brummelen, *The Mathematics of the Heavens and the Earth*, Princeton NJ: Princeton University Press, 2009.

101. Glen Van Brummelen, *Heavenly Mathematics: The Forgotten Art of Spherical Trigonometry*, Princeton NJ: Princeton University Press, 2012.

102. Oswald Veblen and John Wesley Young, *Projective Geometry*, Boston, MA: Ginn, 1910.

103. François Viète, *Apollonius Gallus. Seu, Exsuscitata Apolloni Pergæi Geometria*, in Frans van Schooten (ed.) *Francisci Vietae Opera mathematica*. Lugduni Batavorum: Ex officina B. et A. Elzeviriorum, 1646, 325–346.

104. K. G. C. von Staudt, *Geometrie der Lage*, Nürnberg: Bauer und Raspe, 1847.

105. K. G. C. von Staudt, *Beiträge zur Geometrie der Lage*, in three volumes, Nürnberg: Bauer und Raspe, 1856–1860.

106. I. M. Yaglom, *Geometric Transformations II*, transl. A. Shields, New Mathematical Library, Random House: New York, 1968.

107. I. M. Yaglom, *Geometric Transformations IV*, transl. A. Shenitzer, New Mathematical Library, Mathematical Association of America: Washington, D. C., 1969.

Index

© The Author(s), under exclusive license to Springer Nature Switzerland AG 2025   203
C. Baltus, *Geometry by Its Transformations*, Compact Textbooks in Mathematics,
https://doi.org/10.1007/978-3-031-72281-3